Between the Coast and the Western Front

A cold winter's scene with men with limbers waiting to collect supplies from a roadside dump.

BETWEEN THE COAST AND THE WESTERN FRONT

TRANSPORTATION AND SUPPLY BEHIND THE TRENCHES

SANDRA GITTINS

Lancers making their way through a war-damaged town, September 1917.

First published 2014

The History Press
The Mill, Brimscombe Port, Stroud, Gloucestershire, GL5 2QG
www.thehistorypress.co.uk

© Sandra Gittins, 2014

The right of Sandra Gittins to be identified as the Author of this work has been asserted in accordance with the Copyright, Designs and Patents Act 1988.

All rights reserved. No part of this book may be reprinted or reproduced or utilised in any form or by any electronic, mechanical or other means, now known or hereafter invented, including photocopying and recording, or in any information storage or retrieval system, without the permission in writing from the Publishers.

British Library Cataloguing in Publication Data.
A catalogue record for this book is available from the British Library.

ISBN 978 0 7509 5843 1

Typeset in Minion and Univers by The History Press
Printed in Great Britain

CONTENTS

Acknowledgements — 6
Introduction — 6
Abbreviations — 7

One — Supplies and Distribution — 9
Two — Medical Services — 19
Three — By Canal and Road — 31
Four — Railways — 49
Five — Other Activities behind the Front — 63

Appendix: Relevant Places — 92
Bibliography — 94
Index — 95

ACKNOWLEDGEMENTS

All the official photographs, postcards, maps and other items reproduced in this book are from my own collection.

I would like to thank *Punch* for giving permission for the Shell adverts to be used. I did contact Shell, but had no reply. I hope they are happy that I have been able to illustrate the contribution they made during the war.

I would also like to thank David Bell of the Imperial War Museum photographic archive for his advice, and Nicola Hunt, Crown Copyright Administrator of the Ministry of Defence, for her help in clarifying copyright and publication permission for the official photographs in this book.

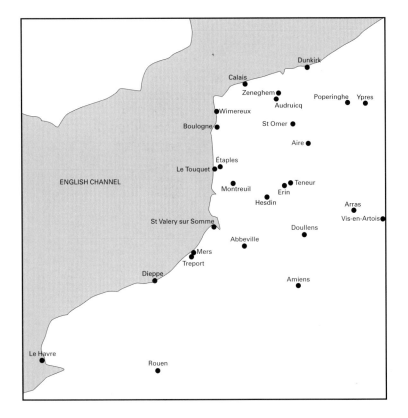

INTRODUCTION

This book aims to take the reader on a trip around the Western Front as known to the British Expeditionary Force, concentrating on the work carried out by various organisations, both military and civilian, in the area stretching from the coast to the front line. This is not intended to be a definitive history of life behind the front line, but a brief outline in the hope it will encourage visitors to the Western Front in these four commemoration years to look beyond the trenches towards the coast – there is much to explore.

Fitted with machine guns and plating, the early armoured cars of the Royal Naval Air Service were Rolls Royces and Mercedes.

ABBREVIATIONS

Stretcher-bearers taking a wounded man to an aid post.

ASC	Army Service Corps
BEF	British Expeditionary Force
CCS	Casualty Clearing Station
CWGC	Commonwealth War Graves Commission
DSO	Distinguished Service Order
FANY	First Aid Nursing Yeomanry
GER	Great Eastern Railway
GHQ	General Headquarters
GWR	Great Western Railway
IWT	Inland Water Transport
LGOC	London General Omnibus Company
LNWR	London & North Western Railway
MMGC	Motor Machine Gun Corps
NCO	Non-Commissioned Officer
NER	North Eastern Railway
OR	Other Ranks
POW	Prisoner of War
QMAAC	Queen Mary's Army Auxiliary Corps
RAC	Royal Automobile Club
RAF	Royal Air Force
RAMC	Royal Army Medical Corps
RE	Royal Engineers
RFC	Royal Flying Corps
RN	Royal Navy
RNAS	Royal Naval Air Service
ROD	Railway Operating Division
VAD	Voluntary Aid Detachment
WAAC	Women's Army Auxiliary Corps
WDLR	War Department Light Railway
WRAF	Women's Royal Air Force
YMCA	Young Men's Christian Association

ONE

SUPPLIES AND DISTRIBUTION

It is useful to know the structure of the British Expeditionary Force (BEF) on the Western Front before getting involved in the story of transport and supplies. The BEF was formed up of five armies, with the Second Army roughly covering the area between Ypres and Hazebrouck, the First Army down to Arras, the Third Army down to the Somme, and the Fourth to the south of the Somme, with Fifth Army in reserve. Each army consisted of two corps and in the corps two divisions. The regulating stations allotted to each army were 1st Army Abbeville, later Boulogne, 2nd Army Riviere Neuve (Calais), 3rd Army Abbeville, 4th Army Romescamps. Daily trains left the regulating stations having travelled from an allotted supply depot with enough stores for two divisions for twenty-four hours, and here additional wagons would be added to the train. The trains were made up of wagons containing the following: mechanical spares, petrol, coal, ordnance, mail, five wagons of oats and four of hay, two wagons of groceries, two of bread and one of meat.

FROM DOCKS FORWARD

At the beginning of the war depots were set up for the holding and distribution of stores for the BEF, but they were temporarily abandoned due to the retreat from Mons and the withdrawal towards Paris following the German advance in 1914. Supplies were diverted from the Channel ports to Nantes and St Nazaire. Once the BEF were able to regain ground and stabilise the position everything was moved back to the Channel ports.

Frozen meat was sent from England but, owing to the limited availability of cold storage, the meat ships were used as temporary stores until purpose-built ones were ready at Boulogne and Le Havre. There was a reserve of preserved meat should the demand for frozen meat be in excess of what was in storage.

It became obvious very quickly that the UK was unable to keep up with supplies to France as well as hold reserves at home for the troops there, so the

A busy scene at the dockside, Calais. Two Foden steam wagons are loaded and ready to take goods away from the docks, while bacon has been unloaded from a ship and is being taken on porters' trolleys to the storehouses in the background. These large wooden stores held goods awaiting distribution to camps and further on to the front. The turnaround of a ship was rapid, and as soon as a ship or store was emptied other ships would arrive with fresh supplies.

government made contracts with countries in the Americas and Australia for regular supplies, which were shipped directly to France where it was inspected before distribution. There was an attempt to hold sufficient reserves in the stores to cover the occasions when supplies were lost or held up by enemy action at sea. It is worth adding that frozen sausages were not a great success!

Bread was an important part of the daily diet and field bakeries were to be found near the front and at the base camps to ensure a regular supply. At the base camps, Aldershot ovens were used to start with but were replaced in 1915 when steam ovens and automatic bread-making machines, made by Baker & Son of Willesdon, were installed at the cellulose factory at Calais, which had the capacity for bulk baking and storage of hundreds of loaves required each day. More of these machines were installed at the other bases and run by the ASC, as were the number of butchers in the area. Bread ration in France was reduced from one and a quarter pounds to one pound, and supplemented by two ounces of rice daily, and two ounces of oatmeal twice weekly on the recommendation of medical staff later in the war.

Condensed milk, dried peas and biscuits were stored in sheds at Calais, together with bacon, which was both popular and readily available. Cheese was also part of Tommy's ration, but reduced when supplies of Dutch cheese ceased, until cheese was available from New Zealand, Australia and Canada.

Sugar was in short supply but the allowance for those in the field was to be no lower than three ounces. To ensure this it had to be reduced to two and a half ounces for those on the front line and one ounce for others.

▶ ASC lorries waiting at a railhead to load up with stores from a supply train from the base supply depots. A variety of stores can be seen in the pile in the foreground.

▼ A view of Boulogne harbour early in 1914. There is a marked difference between the scene in 1914 to later years as all the transport is horse drawn and motor vehicles are almost nonexistent.

▲ ASC lorries are being filled with supplies from a train at a railhead with ASC men checking the paperwork, while in the background troops on a London omnibus are being transported to the front.

➤ ASC bakers are displaying the bread they have baked in some Aldershot ovens at a base camp. A chap in the background is showing a tray of unbaked loaves ready for the oven. The long spade-shaped tools are for getting the trays of bread in and out of the ovens.

Occasional failure of fruit crops reduced the allowance of jam, but in turn increased the sugar allowance.

Potatoes were another staple of the diet, but there was some difficulty in supplying sufficient quantities from home, so additional amounts were sourced from Ireland, Jersey and Italy.

Stacks of crates of tinned goods, resembling small hills, were located near the docks. A Supplies Purchase Department was based in Paris with the job of obtaining local goods, and this department rapidly expanded with branches around Europe. There was a wide variety of goods that the department would obtain, from dried fruits to pigeon food, and strawberries to champagne!

There was no point in having food with no one to cook or organise meals, and as the number of troops increased cooks became in short supply, so cookery schools were established with each army and on the lines of communication. The main schools were at Rouen, Boulogne, Calais, Étaples and Le Havre, and in all some 25,277 men were trained in the art of cooking.

It is also worth noting that the diets of the colonial troops were catered for.

PORTS

The main ports used by the BEF were Dunkirk, Calais, Boulogne, Le Havre and Dieppe, together with Rouen on the River Seine and to a lesser degree St Valery sur Somme.

Sea ports could be closed for days for fear of the loss of shipping when German submarines were patrolling the Channel, but by the end of the war the ports had handled 5,269,302 tons of ammunition, 5,910,000 tons of hay and

oats, and 3,713,208 tons of general supplies. The grand total of all supplies amounted to 27,566,245 tons.

COAL

Coal was shipped from Britain to the north Channel ports, but to save on rail transportation on the Western Front arrangements were made whereby British coal was shipped directly to Le Havre and Rouen for use in Paris, and in turn the BEF was allotted coal from the mines in the Pas de Calais region.

POSTAL SERVICE

This was a vital service for those anxiously awaiting news of loved ones at the front, and they, for the most part, found comfort from the receipt of letters and parcels; a chance for their minds to escape to the familiar things they had left behind. Served by the army base post offices at Le Havre, Boulogne and Calais, the Royal Engineers (RE) Postal Section (special reserve) peace time strength was 300 officers and men, all of whom were General Post Office employees, and it was soon found they had too few men to deal with the workload as demand increased and the men were required elsewhere. Postal workers at home were asked to volunteer and an advert was placed in the Post Office Circular calling for female sorters, clerks etc. with a minimum of four months' experience and aged between twenty and forty, to serve for twelve months or the duration of the war. Those with husbands serving in France were not eligible to apply as the Army Post Office processed the particulars and belongings of men who were wounded or lost, and it was therefore a sensitive

◀ The supplies have arrived and the cooks are getting ready to prepare a meal in the field. Some men are chopping wood for the fire, while others are cutting up a joint of meat. A box of Huntly and Palmers biscuits can be seen, together with a range of cooking vessels and what looks like a large stoneware jar that normally held the rum ration.

▲ ASC men are preparing a field kitchen with food and hot drinks to be taken either on a march with troops or up the line to the front where hungry troops are waiting for a hot meal.

> A not very appetising scene of the bulk delivery of bread at a camp and, judging by the quantity, the bread was probably from the large army bakery at a base, possibly Calais.

▼ A postcard of a sketch showing the ASC travelling field kitchen being pulled by two horses. The hot section of the kitchen can be seen at the back, complete with oven and hot water and tap.

position to hold. The women were to be enlisted in the Women's Army Auxiliary Corps (WAAC) and the first draft arrived in France in May 1917.

The women who went to France were relatively young and lacking in any long-term experience in the Post Office at home, and they soon found out that life in the Army Post Office was to be far more intense. They had to quickly get used to a routine that wasn't too unfamiliar, but more varied than the specific departments they were used to back home. Dealing with postal orders, telegrams, currency exchange and the intricacies of the army censorship system had to be learned by all of them. Working hours were long, usually from 7 a.m. to 11 p.m., and hazardous when enemy aircraft decided to visit the neighbourhood.

Sometimes a meal was eaten wherever was available and these NCOs are tucking into some food amongst the ruins. The chap at the back appears to have enough food for two! These members of a pioneer battalion have obviously seen hard times during their work, judging by the condition of their uniforms, and the sergeant on the left is in desperate need of new boots.

When the front line moved the structures left were put to good use. At this old support line at Kemmel a meal is being prepared with cooking pots in evidence, as is a box of corned beef and one box marked 'peach'.

Sorting the mail at a divisional field post office with ASC lorries ready to take the mail bags onwards to a railhead.

The basic route of correspondence was that a letter written in the field would be handed to a junior officer who would censor and stamp the letter. This would go to the field post office, which was portable and travelled with the division or brigade, and took the form of a black metal box, its location indicated by a red and white flag. All the post would arrive at a railhead before going on to the Army Post Office at a Channel port for shipment to Britain, usually via Southampton or Folkstone, and could take as little as twelve hours or up to three days to be delivered, depending on any action taking place at the time.

The Christmas period was always busy and in the winter of 1917 around 19,000 mailbags travelled across the Channel to France each day, peaking to half a million bags prior to Christmas, which equates to 100 trains of 30 trucks in length delivering to railheads, and for the mail to reach the various field post offices required 6,000 lorry loads.

Troops were always writing home asking for clothes, food, toiletries etc. to be sent out, and the Army Post Office dealt with it all, from a pair of gloves to a Fortnum & Masons hamper.

▶ The Field Service postcard was a mass-produced card and allowed the soldier to send a very brief message by crossing out any sentence that wasn't applicable. Brief it might have been, but if a soldier hadn't been able to write for a while this card would have been reassuring to the receiver.

▶▶ There were many artists with the British forces on the Western Front and they turned their hands to producing Christmas cards. There were many comical ones from both land-based and air force personnel, but this one is the largest card produced, measuring 25cm by 16cm. This watercolour drawing by Percy Vignale tells the story of the 48th Division up to Christmas 1917.

▶ A field post office of the Indian Corps with a curious crowd and gendarme looking on. The original caption of this photo states that the officer on the right was killed the following day, unfortunately his name is not known.

▶▶ On a cold winter's day the arrival of the Christmas post brought some cheer to the men, and it appears that a cat is eager to learn if there is any post for him!

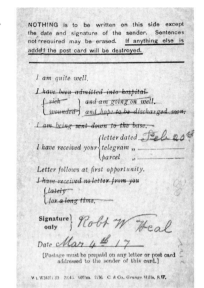

SUPPLIES AND DISTRIBUTION

TWO

MEDICAL SERVICES

MEDICAL

The inevitable consequences of war are wounds and illness, and here is a brief description of the care taken of the soldier from the front line to the hospitals on the coast.

The wounded or sick soldier progressed down a chain starting with walking, or being stretchered, to a regimental aid post for first aid before moving on to a forward dressing station (main or advanced), where the patient had his condition evaluated, his wound dressed or treatment initiated so as to be in a stable condition to go to a casualty clearing station (CCS), or if no other treatment was necessary the man was returned to his unit. Dressing stations were the furthest forward of the Royal Army Medical Corps (RAMC) units. Two forward dressing stations and a main dressings station, usually about a mile further back, was the 'field ambulance' – not to be confused with the motor vehicle.

Dressing stations were makeshift arrangements housed wherever a certain amount of protection from shelling could be provided, although by no means guaranteed.

Depending on the condition of the patient, and the location of the CCS, the patient might walk, be taken in a motor ambulance or other vehicle, including horse-drawn, or by light railway. CCS were large, usually tented, medical facilities located near to a good transportation link of road or rail, and sometimes canal. Here all the emergency treatment was carried out to stabilise the patient, including amputations, the removal of shrapnel and closure of wounds in the operating theatre. Mostly there would be a group of CCS in one place, and with a capacity of around 1,000 patients at any one time, these CCS would be inundated during a battle when hundreds of wounded would pour in. Staff at CCS would move patients in advance when it was known a major action was about to take place to free up the beds.

It is an unfortunate fact that many didn't survive this far, as is evident at the Commonwealth War Grave Commission (CWGC) cemeteries located at old dressings stations and CCS.

The next move for the patient was to the base hospitals along the Channel coast and at Rouen. The favoured means of transportation was by ambulance train, with hospital barges used for patients needing a smooth journey due to their injuries. There were two types of base hospital. The stationary hospital, which

Monchy dressing station of the 44th Field Ambulance Royal Army Medical Corps. A stretcher-bearer is sitting with a wounded soldier while the doctor is attending to the leg wound of a corporal, who is being given a cigarette.

tended to be a specialist unit, for example for gas and neurasthenia, and as isolation hospitals for infective cases. The general hospital was often housed in large hotels and all the latest equipment would be available, with a department for X-ray, bacteriology, dispensary and hygiene. Blood transfusion, still its infancy in Britain as far as technique was concerned, was carried out in the latter years of the war.

Patients who required further treatment, surgery, specialist care or rehabilitation were sent back by sea to Britain.

Women were well represented on the Western Front, with nurses being probably the best known, with the Territorial Force Nurse Service, Queen Alexandra's Imperial Nursing Service, the Voluntary Aid Detachment (VAD) and the First Aid Nursing Yeomanry (FANY). Not all in these services were trained, and those that were varied from fully qualified to volunteers trained by the Red Cross and St John Ambulance. Many in the FANY or VAD were ambulance drivers and cooks, but all the women, whether qualified or not, worked hard under trying conditions and often turned their hands to whatever tasks needed to be done.

Mairi Chisholm and Elsie Knocker were the most photographed women of the war and were greatly decorated for working at the front, contrary to regulations. They were dispatch riders at the beginning of the war, having become friends while competing in pre-war motorcycle and sidecar trials. They joined the Flying Ambulance to aid Belgians, but were aware they could do more by tending to the wounded near the front, and set up their own dressing station in a cellar north of Ypres at Pervyse. They succumbed to

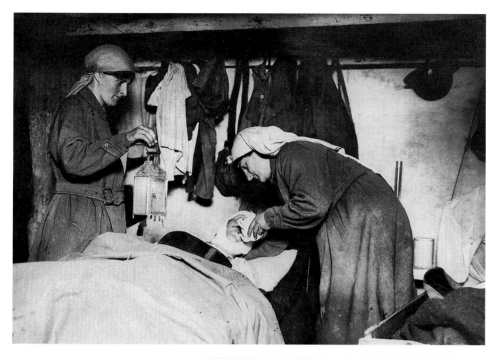

the effects of air and gas attacks in 1918 and spent the remaining time of the war in the Women's Royal Air Force (WRAF).

Dentists were also sent over to the Western Front after an incident in October 1914 when Douglas Haig evidently suffered from toothache. At the time there were no dentists available in the army to treat him and French dentist, Auguste Charles Valadier, had to be sent for from Paris. After this episode, twelve dentists were stationed at CCS from November 1914 and their numbers grew on the Western Front as the war progressed.

▲ The Women of Pervyse; Mairi Chisholm and Elsie Knocker tending to the wounded in their dressings station in the town. The room was far from ideal, but the women were greatly admired for the work they did under trying conditions.

◄ An advanced dressing station in the basement of a shelled building during a quiet time in the summer, judging by the shorts worn by the officer on the left.

➤ Wounded being transported by a horse-drawn tramway on a very uneven line. Note the soldier in a dugout that extends under the line.

▲ The horse-drawn trolleys would have taken ammunition and stores up to the front line and brought back the wounded on its return trip.

◄ Ambulances queue up to take the wounded in a captured village. Many men are lying in their stretchers on the floor. A two-wheeled stretcher can be seen in amongst the wounded.

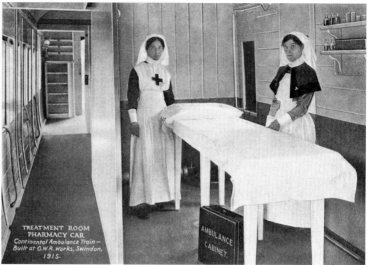

▲ Once the casualties had been dealt with at the Casualty Clearing Station they were transferred to an ambulance train. While on the train they would be continuously cared for until they arrived at the railway station near to a hospital. Here at Le Treport ambulances wait to take the patients on to the hospital.

▶ The pharmacy car contained all that was needed surgically, from the drugs and dressings to a treatment room and operating table, for when an emergency situation occurred that could not wait for the patient to reach hospital.

In 2013 a 1913 Rolls Royce Silver Ghost London to Edinburgh tourer was put up for sale. Sold by its first owner in 1915, it was bought by Haig's dentist Valadier. He was a French-American with an interest in maxillofacial and reconstructive surgery of facial injuries, and his work would help revolutionise the treatment of those injuries during the war. He was said to have been based around Boulogne with his Silver Ghost, which he had modified to be fully equipped with dental chair, drills and other surgical instruments.

Other mobile dental surgeries were available, with the first one made by the engineering department of the Royal Automobile Club (RAC).

HOSPITALS

Hospitals extended all along the coast from Calais, through Wimereux, Boulogne and southwards, with a large concentration at Étaples, on to Le Touquet, Harfeur Le Havre and Dieppe. Inland they could be found at Abbeville, with the largest number at Rouen. All the hospitals were situated to be close to ports and railways.

At Le Treport the Hotel Trianon was home to the General Hospital No 3. Perched high on top of cliffs, and newly built, the hotel was partly used for medical purposes with a large tented area on the extensive grounds. Owing to its prominent position and ease of recognition when approaching from the sea, the Germans destroyed the hotel in 1942.

AMBULANCES

The first motor ambulances available were of a non-standard type with cars being donated with alterations made to the bodies, but this caused problems when it came to repairs and spares for so many different types of vehicle. So the British Red Cross drew up standards to be followed, which meant that bodies could be built and dispatched by manufacturers ready to use. Manufactures included Rolls Royce, Austin, Vauxhall and Morris, but the favoured amongst medical officers was the Ford Model T, with its good suspension and 2.85 litre engine (20hp). Its only drawback was its short wheelbase, meaning stretchers were too long, so holes were drilled into the tailboard to accommodate the stretcher poles! Originally produced in Canada, they were manufactured in Manchester from 1915 and 2,645 rolled off the line by the end of the war.

The ambulance exhaust system ran from the engine into the rear seating area, where it split into two pipes;

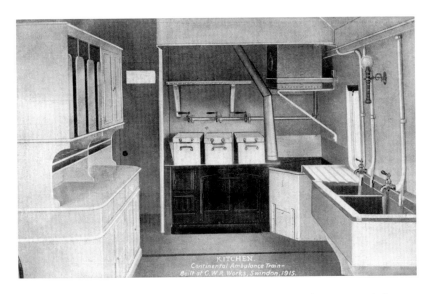

The ambulance trains were equipped with everything that was needed, even down to a cooking stove and running water.

The interior of a ward car of an ambulance train constructed by the Great Western Railway for Continental use. It was obviously difficult to attend to a patient on the top bed, and a wooden step can be seen on the floor by the nurses, presumably for just that task.

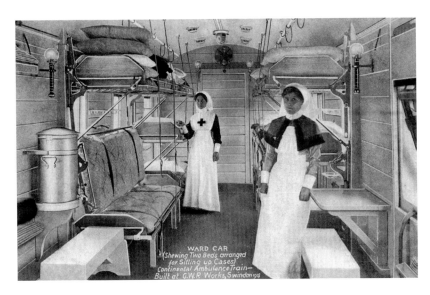

Here the nurses are demonstrating how the middle bed can be dropped down to form the back of a seat for casualties not needing to lie down. This adaptable three-tier system was the innovation of Great Western Railway Carriage Manager F.W. Marillier.

British volunteer workers at a relief station of the French Red Cross. The picture gives good detail of the interior of an ambulance.

A horse-drawn ambulance arrives at a base hospital with the wounded. Soldiers and officers help take the patients to the wards. Ambulances had two roles at the base hospitals: to take the wounded to the hospitals and take those who did not survive to the cemetery.

one of each ran on either side below the seats before exiting at the rear. One assumes that regular checks were made of the exhaust to make sure there were no leaks. This primitive heating system was in place to minimise shock to the casualty; hot-water bottles were also used.

Motor ambulances were able to carry six or eight patients sitting, or four lying, and were driven by men of the Army Service Corps and women of the VAD, FANY and individuals who volunteered.

Many ambulances were donated or sponsored by charitable trusts.

AMBULANCE TRAINS

At the outbreak of war it was not envisaged that there would be a need for Britain to send ambulance trains abroad, as the French authorities were to see to the transportation of troops and the wounded. The French used their *trains sanitaires*, made up of goods trucks and other coaches, which were by no means ideal but they had a carrying capacity 396 patients. They used the Brechot-Despres-Amelines apparatus that allowed three stretchers to be suspended between two upturned U-shaped frames. Due to the great losses during the German advance in the early months and because of the difficulties in obtaining extra trucks, casualties had to be carried in dirty trucks with straw on the floor. This was not to be the last time these conditions were to be encountered, for during the German Spring Advances of 1918 casualties were quickly evacuated in trucks such as these.

The first contingent of RAMC ambulance personnel had arrived in Amiens on 15 August 1914 and took over the goods trucks and coaches from the French and made up the first three official ambulance trains. Everything was cleaned and medical stores and provisions installed. It wasn't perfect and in some cases the train commander had to scour the local shops for items needed, such as blankets, but it was the best that could be done at the time. These trains were put into service 24–26 August, followed by a forth train on the 31 August. The main trouble with these trains was the lack of interconnecting passages between the coaches, so orderlies had to walk along the footboards on the outside of the carriages at great risk to themselves, especially when the train was in motion.

At the end of September the War Office requested that the Railway Executive Committee send a train as soon as they could for use by British troops, using the ambulance train stock available at home. A member of the committee was sent to France to make sure there would be no technical problems.

From August to December ambulance train Nos 1–11 were made up of French rolling stock, and the first to be supplied by British railway companies in November was No. 12, from the Great Eastern (GER) and the London & North Western Railways (L&NWR). All the early trains were well used and covered great distances, and, despite their shortcomings, were well equipped and staffed.

Some of the first English nurses to arrive in France in 1914 can be seen here putting up their tents, which was the usual accommodation for medical personnel. A bulldog has travelled out with them.

In total, thirty ambulance trains were sent out from Britain and continued to be numbered, loosely, in the order of when they first went into service, although there was no number thirteen, for obvious reasons, and no number forty.

Some of the trains were donated, such as No.14, the 'Queen Mary's Ambulance Train', donated by Lord and Lady Michelham. Princess Christian generously gave money for No. 15, the Princess Christian Ambulance Train, and the United Kingdom Flour Millers Association paid for Nos 16 and 17.

An Ambulance Train Committee had been set up in France and taking the advice from medical officers, standardised trains were constructed; Numbers 26–39 and 41–43, with the latter ones having greater capacity for stretcher cases, and were laid out in the following order of lettered coaches:

S: break van for use as infectious ward, living accommodation for guard and attendants, and accommodation for six lying cases.

G: staff car, with sleeping for the medical officers and nurses, mess room, hot water system and shower, and stove.

▲ The 300-bed Hotel Trianon sat on the cliffs above Le Treport. This grand hotel was turned over to the British and became General Hospital No. 3; in the grounds can be seen the tented camp. The hotel was destroyed in 1942 by the occupying Germans concerned that its prominent cliff-top position would be a landmark for the RAF.

▶ Further along the cliff, but still within the grounds of Hotel Trianon, was No. 2 Canadian General Hospital. Only part of this large hospital is illustrated here.

A: kitchen car with an army range, accommodation for three cooks. Also a room for sick officers.

B, C, D, & E: ward cars of thirty-six beds each with toilets and washing area.

F: pharmacy car. For the pharmacy and an area for an operating table. The doors to this carriage allowed stretchers to be taken straight into this area from outside.

L, M, N & O: ward cars of thirty-six beds each with toilets and washing area.

P: divided into compartments for fifty-six sitting casualties with an upper berth for more serious cases.

H: kitchen with accommodation for NCOs and privates.

R: sleeping compartment for other ranks (ORs), also for storing packs and equipment.

T: break van and store, including a meat safe.

The coaches were heated and lit by electric, and there were 300-gallon water tanks in the roof of each coach.

The trains supplied by British railway companies were as follows:

Railway Company	Train number
GER & LNWR	12
GWR	16, 18, 19, 26, 27, 33, 39, 43
GER	17, 20, 28, 36
LBSC & LNWR	14
B'ham Railway & Carriage	15
LNWR	21, 22, 30, 31, 32, 41
Caledonian	23
LB &SC	25
London	38
Midland	34
L&SW	35
NER	37

The carrying capacity varied from train to train, from 262 to 482, made up of sitting and lying casualties. Ambulance trains Nos 3–31 carried 22,709 casualties on the first four days of the Battle of the Somme to the hospitals at the coast.

Together with these trains a further eight temporary trains were in use in France with even numbers between 102 to 118, and they carried another 10,683 casualties during the same period.

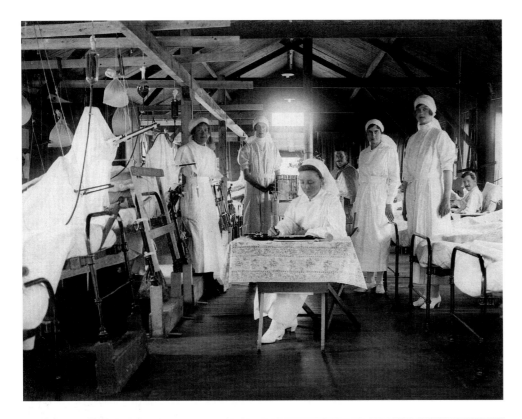

◄ The interior of a ward in the Duchess of Sutherlands Hospital (No. 9 British Red Cross), Calais. Although the hospitals always maintained a clean environment, the nurses' uniforms and bed linen are exceptionally clean owing to extra effort being made for a visit by the king. The patients in the beds are in traction due to what looks like badly fractured legs. The three dark containers hanging from the rafters with lines attached could be blood transfusions, a relatively new form of treatment with the British Army.

◄ A hearty farewell to a group of cheerful wounded soldiers on an ambulance train bound for a hospital ship to Blighty. The majority seem to have head injuries.

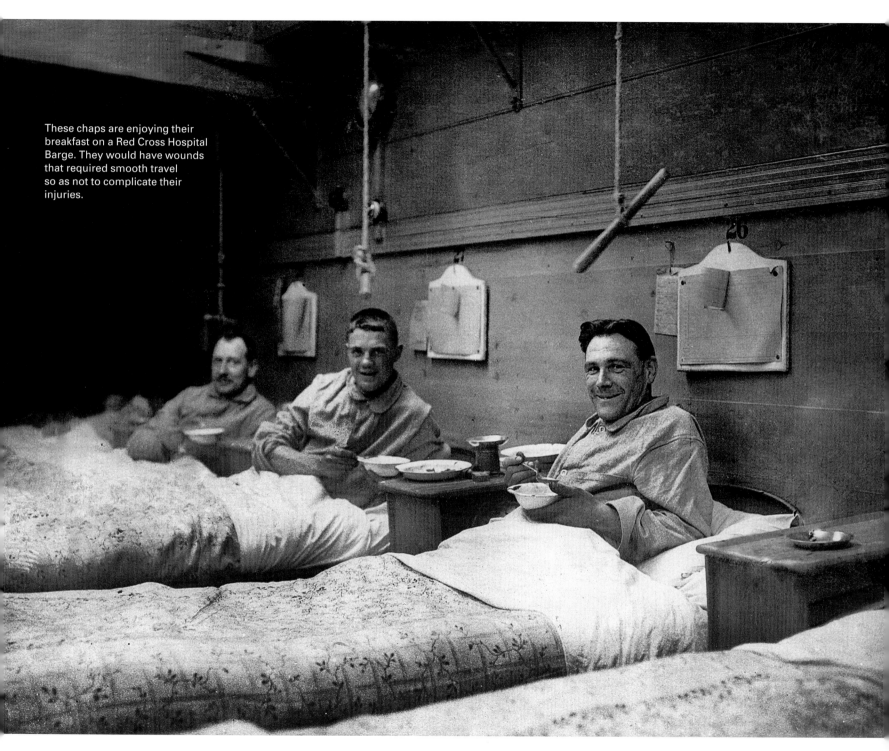

These chaps are enjoying their breakfast on a Red Cross Hospital Barge. They would have wounds that required smooth travel so as not to complicate their injuries.

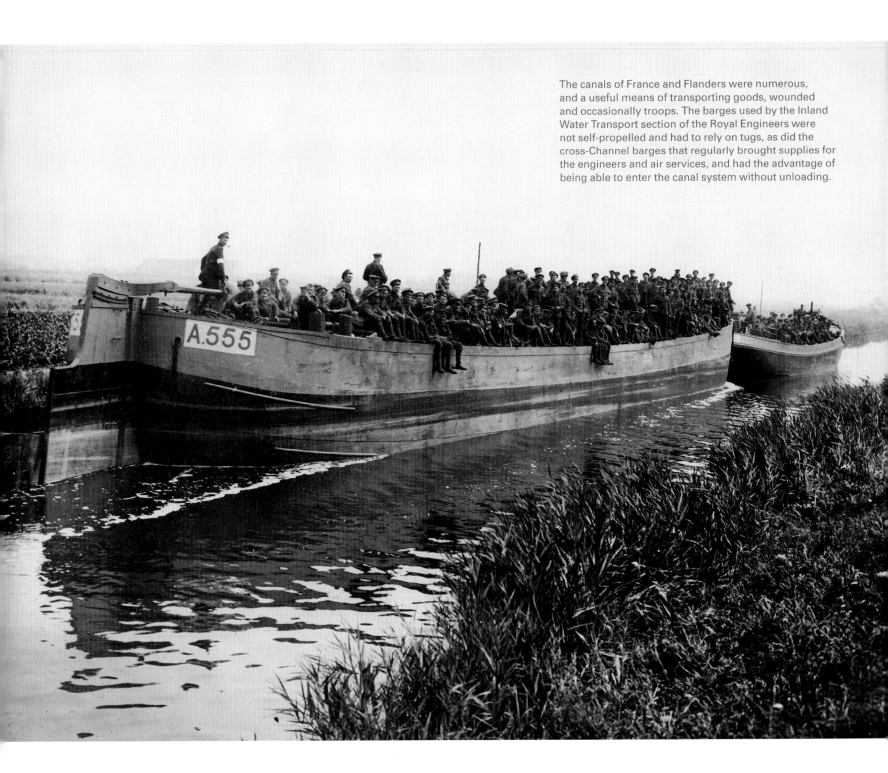

The canals of France and Flanders were numerous, and a useful means of transporting goods, wounded and occasionally troops. The barges used by the Inland Water Transport section of the Royal Engineers were not self-propelled and had to rely on tugs, as did the cross-Channel barges that regularly brought supplies for the engineers and air services, and had the advantage of being able to enter the canal system without unloading.

THREE

BY CANAL AND ROAD

Offloading the wounded from a hospital barge, possibly at Rouen, where ambulances waited to take them to hospital. Judging by the spectators on the quay and lining the bridge, this was a source of fascination to the locals.

CANALS

The extensive network of canals in the Western Front region were put to good use during the war in various ways, as they formed a byway from the Channel ports into the countryside. Troops, goods and spares of every description were transported by barge, but the transportation of the wounded called for modifications to the barges.

There was a great advantage to using barges as their smooth passage allowed for a more comfortable journey, especially for those whose wounds would have made travelling in a motor ambulance, or even a train, unbearably painful.

The disadvantages of using canals was that they, like the main railways, were pre-war routes, mapped and well known to the Germans. Although towed behind tugs, the slow speed of the barges made them easy targets, especially at night as lights had to be used for navigation.

If a canal was close to a CCS then barges could be used as additional bed space for patients. The interiors of the barges were divided to allow the maximum use of space, which included a kitchen, scullery, staff quarters and the central section laid out like a ward, albeit smaller, with fifteen beds lining each side. A hand-cranked lift allowed patients to be lowered down from the deck and although the barges were able to power lights, there was a removable section in the roof to allow light and air into the ward.

VEHICLES

The majority of vehicles on the Western Front were under the control of the ASC, with the Royal Flying

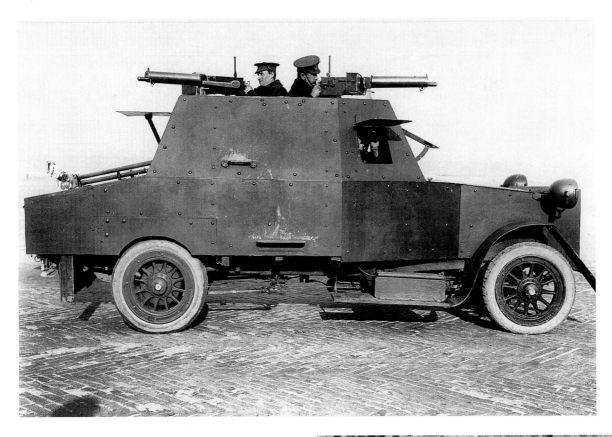

◀ This Royal Naval Air Service armoured car is an early example as the photo was passed by the censor on 2 November 1914. This armoured vehicle was armed with two Vickers Maxim guns.

▼ Samson was greatly revered by the French for his operations against the invading Germans with his armoured cars. Here Samson is photographed with fellow flyers of the Royal Naval Air Service and Royal Flying Corps in Amiens in 1915 prior to him moving to the Dardanelles in March.

Corp (RFC), Royal Air Force (RAF) and Royal Naval Air Service (RNAS) having their own vehicles.

Most of the vehicles owned by Pickfords removal company were requisitioned by the War Office in 1914 and were immediately put to use, as were London omnibuses as these vehicles came under the Voluntary Subsidy Scheme. Under this scheme, help was given for the purchase of vehicles and had to be properly maintained, on the understanding that in the event of war they would be handed over to the authorities.

Private cars, and in some case their owners, were used extensively, mostly for the use of transporting officers. Twenty-five members of the RAC went with their cars to perform these duties.

To counter the, at times, appalling road conditions and early tyres' lack of grip, Dunlop introduced steel

▶ The petrol companies were quick to advertise the part they played in supplying fuel. These Shell adverts explain to the public the importance of petrol-driven engines in wartime. Other companies, such as Pratts, ran similar adverts extolling the virtues of their own fuel and the benefits of a superior product in ensuring the Allies had the best to carry out a victorious campaign. (*Punch* magazine)

▼ Road conditions could be appalling, especially after a winter thaw, as this scene illustrates. The car is axle-deep in mud and it looks as though the driver is revving the engine to within an inch of its life, while others attempt to push, and another car is trying to tow it out of the mud.

studded covers during the war. Fitted to the off front and near back wheel, this configuration was considered the most effective solution to variable conditions. There was a great deal of road building and repair throughout the war.

Many motorcycles were requisitioned and enthusiastic owners went as well, as their skill in riding a motorcycle was not one easily taught.

Tractors were slow but very capable in transporting heavy items such as artillery pieces, relieving horses of the task except over rough ground, although four-wheel drive tractors were available such as Jeffrey and Nash.

By November 1918 there were 26,809 lorries, 708 mobile workshops, 725 steam wagons, 788 Caterpillar tractors, 5,137 cars and over 14,000 motorcycles (with and without sidecars).

Two women of the Women's Army Auxiliary Corps dispatch riders. This is an unusual photo as few exist of female motorcyclists on the Western Front. Those in the Women's Army Auxiliary Corps were drivers, clerical workers, or on household duties. Some were waitresses at general headquarters. These two certainly look as if they are enjoying their job.

Two dispatch riders having a break from their work while resting on their Douglas motorcycles. Because of the nature of the work, delivering messages at high speed over sometimes rough ground, dispatch riders were mostly pre-war motorcycle enthusiasts, and their experience did much to ensure the success and safety of the DRs (dispatch riders).

Motorcycle repairs being carried out in a makeshift garage in France. There are certainly plenty of motorbikes for the mechanics to work on, all of which belong to the Indian Corps.

PETROL

The War Office had contracts with UK petrol companies and required 250,000 gallons a month to start with, increasing to 10,000,000 by August 1918. All the petrol supplied to the Western Front from the UK was shipped from Portishead in cans, with 8,000 cans being filled per day at Portishead. The cans were packed in wood cases for transportation, but a shortage of wood was the cause of the stoppage of can manufacture for a time. Fifty-gallon drums were available, but the preference was for two- and four-gallon cans.

Although cans were easy to handle and stack, this type of container resulted in spillage and waste when decanting, but the cans themselves were reused for a number of purposes, including carrying drinking water – though such water always had a telltale taint of petrol.

Early in 1916, consumption rose to 2,000,000 gallons per month. Landing such a large quantity of cans at the 'petrol discharging' berths at the French docks proved problematic; the answer was to build filling installations in France. The source of supply for petrol switched from the UK to the USA and was shipped direct from the States in petrol tankers to Calais and Rouen, where petrol storage depots had been built, but aviation spirit (possibly a higher octane) continued to be sent from Portishead in cans until the spring of 1918.

It wasn't until the summer of 1917 that bulk distribution was available with the arrival of railway tanks, where the fuel was taken by train to a railhead and pumped into an ASC petrol tank lorry and decanted into cans further on where it was needed.

The petrol company Shell set up storage tanks at aerodromes and railway points. Trains left Rouen with petrol to fill tanks at sidings, which was then either delivered by lorries or underground pipes to the aerodrome tanks. These in turn filled overhead tanks enabling aeroplanes to be filled using gravity.

ARMOURED CARS

The use of armoured cars within the British forces came about out of necessity. Charles Rumney Samson, Officer Commanding No. 3 Squadron RNAS, was based at Dunkirk at the beginning of the war. He was to provide support to ground forces on the French border and into Belgium, but unfortunately there was a shortage of

The Motor Machine Gun Corps was originally a service attached to the Royal Field Artillery and made up of motorcycle enthusiasts. They were of great use during the opening stages of the war, but once the element of speed and surprise had ended and taken over by static trench warfare they were of little use, and many of the men were assigned to regular machine gun sections, but once the army was on the move again in 1918 the usefulness of the Motor Machine Gun Corps was fully realised.

▶ Men of the ASC in a group photo with the lorries in their charge.

▼ The task of spotting enemy aeroplanes and shooting at them was a new operation in this war. Here two observers with telescopes spot the aeroplane, while a range finder gives accurate instructions to the aimer of the gun mounted on the back of a Thornycroft 3-ton lorry. Early examples of anti-aircraft guns were ineffective, but later improvements to weapons and shells made operations more successful.

105. Anti-aircraft Gunners "spotting" a Hun Plane
Official Photograph—Crown Copyright reserved   "Daily Mail" War Pictures

available aircraft. Samson, a pilot, pioneer and innovator (he was the first pilot to take off from a ship) was determined to continue the patrols and was inspired by armoured cars used by the Belgians. However, the only cars available to him were those that some of the men had taken with them to France. He fitted a Maxim gun on his own car and proved how efficient it could be when he ambushed a German vehicle at Cassel on 4 September 1914. This first armoured car squadron consisted of nine men, a Rolls Royce and a Mercedes. Each car was fitted with rudimentary armour supplied by a shipbuilder at Dunkirk and one rearward-firing machine gun.

Armoured lorries were added to the cars and this fleet of vehicles was the beginnings of the RNAS Armoured Car Section. This section carried out impressive work in between Dunkirk and Antwerp, and inland towards German-occupied territory.

The sections title changed to the RN Armoured Car Division when there were twenty squadrons, but

this was short-lived on the Western Front due to the stalemate that ensued with trench warfare, and the division was disbanded in the summer of 1915.

Most of the post-war Bentley racing drivers, known as the 'Bentley Boys', played active parts in the war, and one, S.C.H. 'Sammy' Davis, was in Samson's armoured car section. Davis went on to win Le Mans in 1927.

MOTORCYCLES

Like the Samson armoured cars, a niche was found for motorcycles and dispatch riders during late 1914, and in the spring of 1915 Vickers machine guns were mounted to sidecar frames that were useful during the early phases of the war when there was a certain amount of movement. These adapted machines became the Motor Machine Gun Service (MMGS), and were part of the Royal Artillery, with one battery of these MMGS allotted to each division. Although used at the Battle of Loos in September 1915, the subsequent static nature of the war on the Western Front restricted their use, and they really didn't come into their own again until the final advances in 1918. The MMGS, along with the RN Armoured Car Division, was taken over by the Motor Branch of the Machine Gun Corps.

The motorcycle dispatch riders were in use throughout the war; their ability to transport small, urgent items on congested or damaged roads proved invaluable.

Motorcycle production increased to cope with demand, and many motorcycles were acquired for government use from private owners. Many of the men of the MMGS and dispatch riders were enthusiasts and motorcycle club members.

◀ Solid rubber tyres may have had the advantage of no punctures, but when a lorry tyre needed replacing a hydraulic press had to be used, whereas inflated tyres on wheels were easily replaced with the pre-war innovation, the 'Warland Dual Rim', which was a detachable rim with an inflated tyre that could be quickly attached to a wheel. This lightweight alternative to a spare wheel was used on many military vehicles including the armoured cars.

▲ A repair workshop, likely to be Rouen, where the men are busy at the benches repairing and making parts for motor vehicles. Many different makes of vehicle were in use so there was a lack of standardisation of spares, but this was a war of recycling and a wrecked vehicle would be used for spares. When a part wasn't available it would be made on the bench, the engineering being much more basic than today.

➤ With snow lying in this newly captured village, the inhabitants welcome the officer and men of a cyclist battalion. The bicycle was useful for reconnaissance and communications work. They were quick and required less maintenance than a horse.

Motorcycles used included Matchless, Premier, Zenith, Enfield, Cyclo and Douglas. The RFC also used Phelon and Moore.

OMNIBUSES

The London omnibus is a well-known First World War transporter, being the double-decker Daimler B-Type, introduced in 1910 by the London General Omnibus Company (LGOC), and powered by 40hp Daimler engine. Over a thousand omnibuses were commandeered by the War Office from the LGOC, a third of its working fleet; the famous 'Ole Bill' was one of them. The first buses to reach France at the beginning of the war were slightly different, being single-deck B-types that were converted to ambulances, but these were apparently lost in the early fighting.

Seventy D-type Daimlers were used by the Naval Division from September 1914, and were non-standard vehicles of the LGOC but of great use to the Marines at Antwerp where most were lost. Those remaining saw use in October 1914 during the First Battle of Ypres.

The first omnibuses to arrive were still in their original liveries, complete with adverts, but their bright red-and-white paintwork was a little too easy to spot by the enemy, and they were quickly painted khaki. The omnibuses were driven by volunteer LGOC drivers who were eventually enlisted into the ASC. Even though they were restricted to a speed limit of 6mph in towns and 10mph on rural roads, their troop-carrying capacity of twenty-five men meant they were invaluable, especially when there was to be a big offensive. In July 1917, when ninety-two omnibuses

The 75hp Holt petrol tractors were employed as gun tractors and although slow moving their advantages were that the petrol engine was less conspicuous to the enemy than steam-driven vehicles, plus its caterpillar tracks made it easier to cross uneven ground.

Built by Foster of Lincoln for Daimler, this 105hp Foster-Daimler petrol tractor is towing a naval gun on a specially designed trailer. The gun, it has been recorded, was being moved to its new position on the coast.

◀ A Foster-Daimler towing a mobile workshop.

▼ A line of omnibuses in Poperinghe in 1915.

▲ Khaki-painted London omnibuses full of troops. The equipment on the back consists of picks and shovels for getting the buses out of mud or clearing obstacles encountered on the roads, with emergency maintenance tools in the boxes. All the glass had been removed and replaced with planks as the glass was easily damaged by the men's packs.

▶ A heavily congested Grand Place Bethune with over a dozen London omnibuses, lorries, ambulances, horses, limbers and equipment.

▶▶ Red Cross vehicles and a London omnibus at La Panne on the Belgian coast amongst the sand dunes, in 1917.

and fifty-three lorries (which were omnibuses with the body replaced with a lorry body on the original chassis and capable of carrying twenty men) were used to move troops up to the front line. With the speed limits came the additional problem of having to carry out movements under the cover of darkness, with no lights showing on roads often covered in mud or damaged by shells.

Conditions for the drivers were uncomfortable in the open cab, often covered in mud and dust in the summer and subjected to freezing conditions in the winter. Canvas was used to try to give some protection to the driver from the cold and a canvas cover was put over the bonnet with straw underneath to keep the engine warm.

As well as the conversion to ambulances and lorries, twelve omnibuses were converted to pigeon lofts for the RE Signals to join the early lofts made from converted general service wagons.

ROADS

Traffic was continuously using the roads, with supplies and men constantly being moved from the bases to the front and back again. In the beginning the French maintained the roads, but soon found they were unable to keep up with demand in the British sector as well as looking after their own sector.

French roads were not robustly constructed and the continuous wear caused by excessive traffic, which they had never been designed to take, meant that during

◀ This official Canadian photo was taken on the Arras to Cambrai road, modern-day D939, at the junction for Haucourt and Eterpigny, looking towards Vis-en-Artois. On this busy dusty road, repairs are being made to a motor car, while an aeroplane, which has probably had to land due to damage or fault, is being taken back to base with a Royal Air Force lorry. Some men have gathered for a rest at a Canadian Chaplain Service coffee stall.

▼ A town on the Western Front bustling with troops waiting to go into action, with ambulances and horse-drawn general service carts filling the road. It is wintertime and many of the soldiers are wearing balaclavas under their helmets.

the winter months the roads were deteriorating. By December 1915 the British needed 3,370 tons of road metal a day for their area, and as there was insufficient supplies of good material, granite was imported from Guernsey and other places, which meant extra work at the ports.

Road circuits were drawn up for each army's local requirements, such as the location of railheads, dumps etc., and traffic was excluded from some roads in order to keep them in good condition for when they would be needed for any large-scale movement.

Maps were produced with colour-coded routes. Ordinary roads were marked red, indicating traffic could travel in both directions; blue roads were for light vehicles and motorcycles. Maps printed for traffic prior to any battle included additional circuits for use by motor ambulances. Few roads near the front line could be used by lorries. There were manned traffic-control points to ensure that everything ran smoothly.

A busy, muddy road near Pilkem in 1917. The troops are marching to their destination, heavily laden, their eyes looking down, picking their way. Officers in their cars squeeze past, with limbers and lorries ahead. An ambulance and limbers are attempting to proceed in the opposite direction. Some of the soldiers are making their way up the bank to their posts. Some sappers are busy digging by a tent, which is probably a medical post for below an ambulance is patiently waiting. Many of the men here would not return.

BY CANAL AND ROAD

◄ When there were large movements of troops by motor vehicles maps were produced showing the direction and type of vehicle permitted to use certain roads to minimise congestion. Here is a good example of traffic on an army-designated one-way road at Abele in September 1917.

Lorries and omnibuses couldn't be parked on soft ground so needed parks with hard standings. These were often miles away from army routes but meant they were out of sight from aeroplanes patrolling the main routes.

To give some idea of traffic numbers, on 22 July 1916 at Fricourt cemetery, 7,300 motor and horse-drawn vehicles passed in twenty-four hours.

◄ The maps shown here are the types produced during the war for road and rail use. Most have routes over-printed on existing maps, such as on the Ordnance Survey maps, or specially printed maps.

A road being made through a devastated war scene.

▲ The aftermath of battle. This stretch of the road between Ypres and Zonnebeke has been cleared of destroyed vehicles and other detritus that has been pushed to the sides of the road. Nothing is left standing in the countryside around. Men can be seen working down the road, making good damage done by shelling.

◄ Engineers and labourers are finishing a new road. Sunken roads such as this afforded some protection from shelling. Although a foundation could be laid of chalk or stone, with small stones on top, here a corduroy or plank road has been laid made of wooden sleepers placed widthways over long planks running the length of the road which levelled out the ground. This type of road was designed to be quickly put down and to reduce the risk of animals and vehicles being stuck in mud or sliding off the road. Horses are carrying ammunition in purpose-made saddle packs.

◀ Royal Engineers constructing a pontoon bridge.

◀ A group posing on the newly constructed pontoon bridge pictured in the above photo. They are probably the engineers who worked on this extremely versatile temporary bridge. The combined weight of the men and car demonstrates the sturdiness of this floating structure.

➤ Shells being loaded onto pontoon boats on the River Scarpe at Blangy on the outskirts of Arras, April 1917. The ground had recently been taken by the British and the boats worked to supply the forward guns with ammunition. With this advance came the opportunity to lay a new 60cm line along the newly held stretch of the river. The Simplex on the line looks to be bringing up fresh ballast.

▼ It was vital that the lines of communication remained open, so it was imperative to repair roads, bridges and railways as soon as possible. Here men of the Royal Engineers and a labour company are attempting to make a track down to a bridge they are constructing, while infantry troops pass by.

◀ The pontoon boats are following the river down towards Fampoux. These boats were usually used for temporary bridging work. Some highlanders are making their way to the front by following the new line.

FOUR

RAILWAYS

◀ Ministry of Munitions 2-8-0 locomotive with the Railway Operating Division No. 1881 (works No. 21858) pictured at the depot at Outreau, Boulogne. The officers and men are posing for what looks like an end of war photo. Two Women's Army Auxiliary Corps are sitting next to the officer and they probably carried out clerical work for the depot. This locomotive was acquired by the LNER (6507).

▼ Map of Arras highlighting the railway lines. Pontoon boats followed the light railway alongside the river, running from square G into square H.

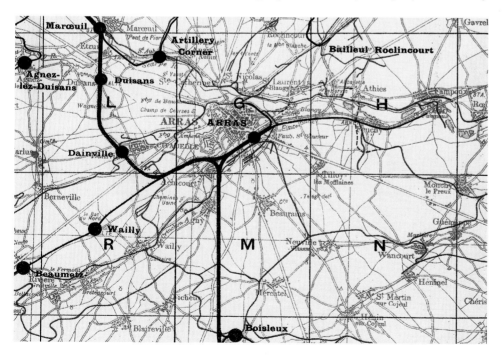

RAILWAYS

At the beginning of the war all the railways on the Western Front were run and maintained by the home countries of France and Belgium, and although one railway company arrived at Le Havre with the BEF on 15 August 1914 there was no work for them to do. This company, No.1 Railway Company (8th Railway Company R.E.), busied itself working with French engineers, but from this slow beginning would end up being one of the many triumphs of the war.

As more and more British troops arrived the number of Railway Transport Officers increased. They worked hard, trying to ensure that the trains designated to the BEF were running as efficiently as possible under the circumstances, but the railways in the area were beginning to feel the strain of such continuous traffic over mostly single lines, and the lack of rolling stock due to the use or loss of French wagons. It was obvious that the situation couldn't continue, and during the first winter the situation was discussed and in January 1915 it was decided to send Railway Operating Division troops (RODs) to France, and they arrived in March.

To give some idea as to the traffic running on the railways in those early months a doctor at St Valery-sur-Somme noted that everyone could hear the whistles of the trains echoing across the bay, especially at night, when train after train travelled on the main line without interruption.

The ROD's troops was made up of units of 270 men. The first five units to arrive in France consisted of three mechanical units for the maintenance of locomotives and two operating units. It was important to have someone in charge of the ROD and Cecil Paget of the Midland Railway was chosen. Like all the men involved on the railways on the Western Front he was a railway man, with a wealth of experience. He was of the opinion that he was there to do 'railway work' and was not the least bit interested in what army rank he was to have, and being perfectly happy with the rank of major; the authorities had to be very persuasive later to give him a higher rank.

The first ROD units worked at the Boulogne docks, with a detachment, based at a sugar factory outside Calais, overhauling twenty-five locomotives. They had to move from the factory during the 'beetroot season' as the factory was needed for processing, so they moved to Caffiers until the new locomotive repair shops were available for use at Audruicq.

Although the ROD men had uniforms they were initially of blue serge, as there was no other material available. This caused much confusion with their fellow army colleagues thinking they were conscientious objectors, so as soon as they could the men got themselves into army service dress.

The first task of the ROD was to learn the Hazebrouck–Ypres line. This was an important line of communication for the British once the Ypres front had become stable, but as the line crossed the border it was worked by the French and Belgians until the beginning of 1915, when the French took control. This route was only single line and inadequate for the work ahead, so a survey was carried out to double the line, which began in April. With more ROD troops arriving, an arrangement was made with the French to take over the running of the line, which they did on 1 November. Civilian passengers and goods continued to be carried on this line, but as the civilians were crossing the border they were not permitted to travel without a permit.

◀ A busy scene at Arras station with a large crowd looking on and mingling with British troops. The lack of damage to the station buildings would suggest this was early in the war.

◀ Troops being taken by standard-gauge train are usually in closed carriages, but here a large number of officers and men, possibly part of a division, are being moved in open trucks. The lower speeds in France would have made the journey less dangerous than it might look at first, but the journey would not have been without risk.

▶ Sandbags were a common sight at the front, or indeed anywhere protection was needed, but how did they get there? Empty sandbags were easily transported, usually from England, being flat canvas forms, but sand was a different matter. Sand was extracted from quarries and taken by rail to a point where the sand was offloaded and the backbreaking, monotonous job of filling the bags was carried out. The filled bags could then be transported to their final destination.

Such was the beginnings of the ROD. There had been a pre-war agreement whereby the French would provide all that was necessary on the railways for the BEF, but in November 1916 the French suspended this agreement, allowing the British to be in control of the lines within their area, and run by the ROD.

When Sir Eric Geddes was sent to the Western Front in 1916 to assess the transportation systems in the area he set into action a great deal of expansion and reorganisation, having calculated the amount of munitions and stores required for the British armies at the front during an advance, and concluded that 300–350 additional locomotives had to be sent to France immediately, together with 20,000 trucks and 1,000 miles of track, plus the relevant maintenance personnel. Only Britain could supply these at short notice. Locomotives and wagons were sent, and lines at home were either closed or double track reduced to single track to meet the need. British locomotives would be supplemented with locos of many nations during the war, and the number of locomotives on the Western Front would peak at 1,376 in October 1918.

Geddes had also stated that with the increased tonnage needed to be carried it was important that light railways be constructed. The standard-gauge (or broad-gauge as was the term used then) railways could only travel so far, with motor vehicles taking the supplies on from the railheads, but with the roads in a bad state of

repair together with congestion, light railways were the most effective means of transportation to, or as near as possible to, the front lines. They had been in existence since before the Battle of the Somme, but they were effectively just tramways as their hauling power was by animal or human, with only six petrol tractors available throughout the front. The benefits of the light railways were that they were easy and quick to install, especially with pre-fabricated sections of track, and, unlike the standard-gauge lines, an army was allowed to more or less lay a line within their area that best suited their needs, without the intervention of GHQ. The light railways were originally under the control of the ROD until they were appointed their own director general.

The railways needed experienced men to build and maintain the tracks, points, signals etc., and as such the only logical place to find these men were from the railway companies at home. The men were first sent to Longmoor for their initiation into army life before being sent abroad. Some of the railway companies formed their own Railway Construction Companies such as the Great Western Railway (GWR) who were able to form three companies, one of which was sent to Egypt, while the other two worked on the Western Front; one of which, the 262nd, did much work in the Ypres area prior to the Battle of Passchendaele, not without loss. All the construction companies travelled to their area of work on a travelling train, with wagons containing all

◄ Railways were not just for transporting troops and supplies but were ideal for positioning railway-mounted guns. These large guns were for long-distance use, and could be moved from a gun spur to a siding in a different area relatively easily to be in the best positions for effective use.

◄ The railway workshops at St Etienne-du-Rouvray, Rouen, would have been a familiar place to most British railway men, except for the variety of locomotives. Here a 2-8-0 Baldwin (USA) is being worked on under the watchful eye of two officers.

› The company office of the 279th Railway Company Royal Engineers somewhere in France. Men and junior Non-Commissioned Officers are standing, while the officer and sergeants are seated, with dogs. A railway company would use railway trucks for their offices and stores which could be moved when necessary.

they needed, such as tools, general stores, tents and an office, and all were assisted by labour companies.

In all, thirty-five companies were raised, mostly in Britain, but two were raised in Boulogne when a call went out for anyone with railway experience already serving on the Western Front.

Three-thousand civilian platelayers and gangers were called for by Geddes in February 1917, such was the urgency for construction work to be carried out. Eight companies were formed from the volunteers and assigned tasks in Flanders. Number 1 Company of the LNWR worked the whole of the time at Audruicq. Unfortunately, the ground was of clay and so urgent was the work that the rails were in use as soon as they were laid, often without ballast, so the rails sunk into the ground and the LNWR men had to re-lay them frustratingly often.

Two men from one of the GWR civilian companies decided to stay and enlisted in Belgium in the 262nd Railway Company; unfortunately they were killed shortly after.

The British Light Railway Companies consisted of seventeen operating companies (seven were raised in France), five companies of train crews, three of workshop and seven forwarding companies. There were also Australian, Canadian and South African companies operating on the Western Front. The number of locomotives working the lines was at its

◀ Engineers are either replacing damaged track or laying new track in the snow on what was land held by the Germans. This would have required experienced railway gangers as well as labourers to carry out the work as quickly as possible, so the line could be used.

▼ A signal box was an important target for enemy artillery and bombers with the delays to trains that would result from an attack such as this. Most of the brickwork has been destroyed, but the levers have stood up to the attack quite well. It has been recorded that this signal box was at Artillery Corner, Arras.

height during the Battle of Pilckem and Passchendaele in 1917, at a total of 546.

It was the duty of the RTO to arrange and oversee all railway traffic, an often hectic task with traffic consisting of ambulance trains, supply and forage trains, engineer and works trains of stone for ballast and other building material, coal, and trains for remounts and reinforcements.

AUDRUICQ

Audruicq was an important RE stores and ammunition depot. The ammunition depot, commanded by Colonel Walter Gordon Wolrige-Gordon, was attacked by the German air force on 20 July 1916 and the depot was bombed. Major Barrington-Ward ROD was awarded the DSO for driving an engine between some sheds while the ammunition was exploding to bring fifteen men to safety.

Captain F.C. Willis used his experience of working with the GWR stores system at home to manage the RE

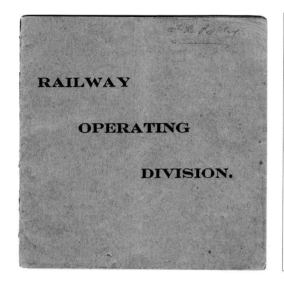

RAILWAY OPERATING DIVISION.

WORKING INSTRUCTIONS.

The following instructions are issued for the general guidance of Officers, Non-commissioned Officers and men of the Railway Operating Division of the British Army in the field.

It is impossible to provide for every contingency which may occur in the working of a railway in the area of military operations.

Circumstances will undoubtedly arise in which the personnel must decide for themselves what should be done and act on their own initiative with no rule to guide them.

They must understand that their first duty is to satisfy the requirements of the Army. Ordinarily, they will receive orders from railway officers only, but in cases of unforeseen circumstances in the neighbourhood of the enemy they may receive orders from any officer. They should ask for such orders in writing and forthwith comply with them, reporting the fact of having received them to the nearest officer of the Railway Operating Division, or the Railway Transport Establishment, without delay.

They are responsible for taking every precaution to prevent accident that their knowledge of railway working suggests.

J. H. TWISS, Brig.-General,
DIRECTOR OF RAILWAYS

G. H. Q., 1st Echelon,
June 10th, 1915.

TRAIN ORDERS.

The line will be divided into sections, and at each block post hand signals will be given to the drivers in accordance with Clause 2, but no driver will be allowed to proceed into a section until he has been furnished with a TRAIN ORDER, indicating either

(A) That the section is clear, or
(B) That the section is occupied, but that the preceding train has left not less than 10 minutes previously.

In the latter event the driver will run " At Caution," i.e., at such a speed as to be able to bring his train to rest at any point on the stretch of line within his view.

The driver must sign this train order and give it up at the end of the section to which it refers.

CANCELLATION OF TRAINS.

When it is found necessary to cancel the passage of a train into a section, the following procedure must be rigidly adhered to:—

The Section Controller must instruct the guard to get from the driver his train order and hand it over to him. When the Section Controller has received the train order from the guard, **and not until then**, he will ring up the post in advance and cancel the train in accordance with the printed instructions in each despatching post.

▲ A booklet produced under the direction of Brigadier-General J.H. Twiss, Director of Railways General Headquarters, for the guidance of all those involved with the running of the standard-gauge railways.

➤ Shells are being pulled along a tramway to the guns. The weight of the shells and the conditions underfoot ensured this was not an easy task for the men pulling the load, but these tramways were almost noiseless and allowed shells to be moved up to the gun positions at night near to the front line.

stores at Audruicq. As well as the REs, Chinese labour was employed in this vast store. It is worth noting that not everything arrived from Britain and often purchases were made locally, like cylinders of oxygen for oxyacetylene welding that were obtained regularly from Usines de l'Air of Calais and transported to Audruicq by RE lorries.

▲ Built at Crewe by the London & North Western Railway, this vehicle was a dual-purpose road and rail vehicle. It was based on a Model T Ford with standard engine and running gear, with a turning plate added so the vehicle could be lifted and turned in the opposite direction when on the rails. Tests had been carried out at Crewe and found that the vehicle capable of pulling 5 tons. Although 132 were made they were not successful as the track encountered near the front was nothing like that used at the tests, and they also had difficulty progressing with heavy loads without assistance.

◀ This atmospheric photo shows that even where the ground is waterlogged the line is in excellent condition. Some chaps are having a brew next to a line-side hut in the background, watching a working party go past on the train. The hut was probably, like at home, for storing tools and guarding the line, and as the hut has a chimney there was probably a fire inside for heat and cooking.

The official title for this photo is 'Short cuts to the line' and illustrates that a narrow-gauge line constructed during the war could take advantage of obstacles in its path. Here the route makes good use of an archway in a war-torn building in Arras in 1918. The soot marks on the arch is evidence the line had been in regular use for a while, although the precarious roof tiles are a hazard!

◀ A snowy scene at Elverdinge, north-east of Ypres, in February 1917. This Hunslet 4-6-0T still has its original markings of the Railway Operating Division. This type of steam locomotive was the most popular, with seventy-five being ordered. Here it is transporting equipment for a work party of cheerful-looking men.

▼ An ammunition train on a light railway near Elverdinge. This photo has the following official number of the Hunslet photo so they were presumably taken at the same time. The winter of 1916–17 was bitterly cold as can be seen here, with rock-hard frozen ground, which caused no end of problems to water supplies, roads and railways.

RAILWAY ACCIDENTS

Some accidents occurred owing to the poor state of the tracks and the line from Doullens to Candas, on the Amiens line, was no exception and had gained a reputation as an accident black spot. On 24 November 1916 a train full of officers and men about to go on leave departed Doullens station. Unfortunately some wagons had broken loose and, owing to the gradients, ran at speed into the leave train. Details are sketchy but according to reports the accident happened in the vicinity of Gezaincourt and eighteen officers and fifty-one men were badly injured owing to crash injuries and burns. It was said that seven bodies were so badly charred that they were unrecognisable.

Close by was No. 29 CCS whose members quickly attended the accident. Captain Charles Hoskyn RAMC went to the aid of a trapped man who had one leg burned off. Hoskyn knew he had no alternative but to amputate the other leg pinned down by debris, but he managed to move the man sufficiently to free him. Sergeant John Orr

Built in April 1917 by the Gloucester Railway Carriage and Wagon Company, this generator car was one of six vehicles that went to make up a workshop train for the 60cm railway. Inside can be seen two petrol-driven Aster engines to drive dynamos, together with a motor-driven air compressor. Exhaust outlet pipes and silencer from the Aster engines can be seen on the roof.

RAMC battled through the flames and heat to rescue men, including a man pinned down by heavy timber beams on his legs. Hoskyn was awarded the Albert Medal and Orr received the Meritorious Service Medal.

Ten burials can be found in Gezaincourt Communal Cemetery Extension from that date, two officers and eight ORs.

TRANSPORTATION OF TANKS BY RAIL

First World War tanks were a great new innovation, but were slow vehicles with a high fuel consumption. They could, therefore, only travel relatively short distances by road, and by doing so caused considerable damage to the surfaces. For these reasons movement by road was not an option, plus these new weapons were secret and any movement was, wherever possible, to be undetectable from the air or general public. The tanks were originally shipped from Portsmouth to Le Havre and offloaded by crane (Boulogne was used when the crane at Le Havre was under repair), and then transported by rail to their final destination.

The first tanks arrived in France in August 1916 and unfortunately little thought was given as to whether stock was available on the French systems capable of dealing with the weight and size of these vehicles. All rolling stock on a railway is governed by a loading gauge, that is its height and width being capable of going under bridges and passing on the tracks. The tanks as they stood, loaded on a wagon, were too wide. The answer was to remove the sponsons (the side protrusions containing the weapons). This also reduced the weight of the tank, but even so there were few trucks capable of carrying such a weight (the lightest tank was 15 tons and heaviest over 28 tons). The main French railway system used was the Nord, which had flat trucks with a 20-ton carrying capability, but they were unable to pass the loading gauge of the Etat system that served Le Havre. The Etat railway had wagons capable of carrying 25 tons and these were used to carry 28-ton tanks. Forty-three of these wagons were used for the movement of tanks for the Battle of Cambrai in 1917 and all except fifteen were severely damaged by the excessive weight carried.

An alternative had to be found as quickly as possible. British wagons used for carrying rails had been sent to France, and they had a 45-ton load capacity, but these were being worked on the railways throughout the Western Front and it took some considerable time to collect sixty of them at Audruicq, where they underwent conversions to the unsupported overhangs to allow for the end-loading of tanks.

There were moves in Britain for the construction of special trucks, but for the meantime the GWR supplied wagons capable to bearing 25-tons until the Railway Executive Committee's tank trucks, or Rectanks,

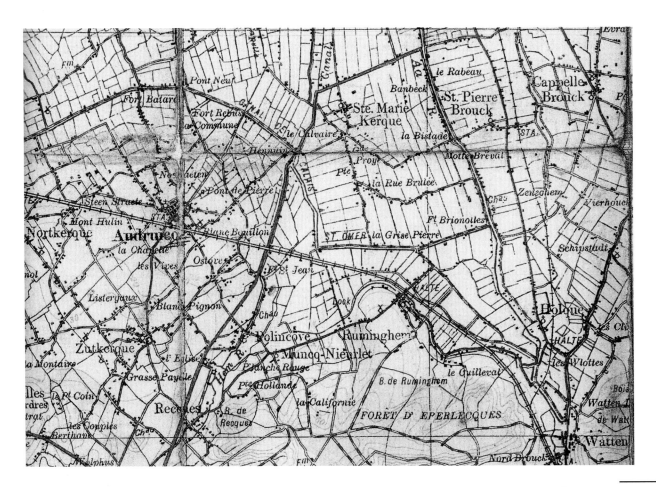 

◀ A 1917 map of Audruicq showing the layout of the Royal Engineers and ammunition depot. Zeneghem can also be seen on this map on the right. It was the site of an ammunition depot and for baulk wood – this place is no longer marked on modern maps.

were available. These Rectanks were capable of carrying 40 tons, and around 400 were in use in France by the end of the war.

The loading and offloading of tanks on wagons was another problem. Large earthworks were constructed by the side of the railway at the tank workshops at Teneur to allow them to be loaded onto the side of a wagon, and an overhead gantry was installed at the tank depot at Erin, but this was a slow process and an easier way had to be found that could be used wherever the tanks were needed. End loading of tanks was the only efficient means, with the tanks driving up a ramp onto the trucks and along, forming a line of tanks. Sleepers were placed over the rails to protect them from the weight while the tanks were being manoeuvred. Many ramps were built at Plateau, with entraining stations at Ruyaulcourt, Heudicourt etc. The ramps were made of sleepers with an angle of 1 in 7, and took twenty men ten hours to construct.

During battle, when a large number of tanks were in use, the ROD had to ensure there were sufficient locomotives available for the task. From 15 to 18 November 1917 nine trains of twelve tanks ran each day (432 tanks), with the corresponding return trains from Ruyaulcourt to Plateau over two weeks in December transported 336 tanks. By 1918 tanks already loaded onto railway wagons were transported from Britain on the cross-Channel ferry from Richborough.

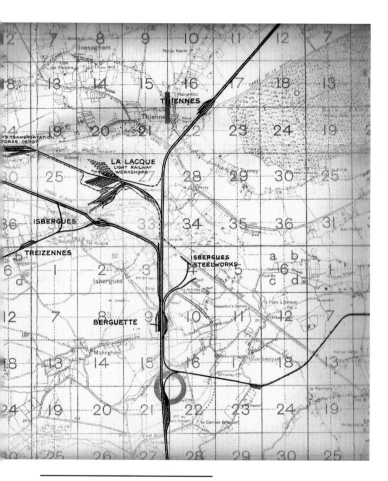

▲ A map showing the location and rails around the light railway workshop at La Lacque.

▶ Map of rail lines around Bethune.

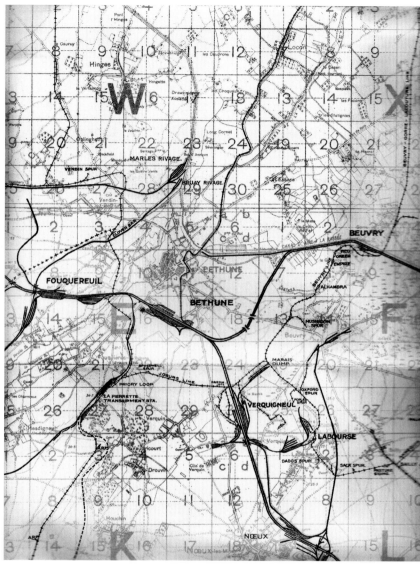

By any means of transport the carriage of ammunition was fraught with danger, either by enemy action or accident. On 30 April 1918, at a refilling yard near Poperinghe, an engine had just been uncoupled from an ammunition train when the second truck burst into flames. Sergeant-Major Alfred Furlonger ordered Corporal Bigland to return to his engine and take the first two trucks away, but when they were a safe distance away from the ammunition dump the burning truck exploded while being uncoupled, with enough force to throw the engine 50 yards, killing three of the men of the Light Railway Companies, RE. They, along with two of their colleagues, were awarded the Albert Medal.

◀ Two Tommys preparing shells. It was a dangerous task, as the notice in the background implies: 'NCOs and men are not allowed on this dump unless they are working here!'

▼ An ammunition dump on the Western Front with shells ready to be transported to the guns near the front line. All the men in the photo would have been needed for the backbreaking work of lifting these heavy shells.

FIVE

OTHER ACTIVITIES BEHIND THE FRONT

FORESTRY AND FARMING

The demand for wood for use in trenches, railway and road maintenance and repair was great. Woods in Britain were under government control and this timber was supplemented with imports from America and Canada, but these were subjected to enemy action at sea. Large quantities were also sourced locally following an agreement with the French government. Timber was processed at timber yards located in forests. The unsuitable wood was burned to produce charcoal in the traditional way.

Land vacated by French and Belgian farmers was readily put to use. Anything that was of use was harvested, from potatoes to hay, with troops and machinery working the land, overseen by the Directorate of Agriculture. A large pig farm was established at Étaples, but there was encouragement to grow vegetables on any available ground, even disused trenches, and keep chickens; anything to supplement supplies. This was a war of salvage and recycling. Degreasing plants were set up at home and at Étaples, Boulogne, Dieppe, Calais, Rouen, St Omer and Le Havre, which extracted fat from waste of the camps. The fat was then processed and glycerine was extracted, which in turn was used into arms propellant for the Ministry of Munitions.

Everything was collected for recycling. Empty ammunition shells of all sizes, clothes, boots, cutlery – in fact, if it could be recycled it was.

WATER

Clean water for men to drink and for medical purposes was in short supply nearer the front and the ASC had motorised water tanks for this purpose. Horses and mules also needed vast amounts of water, and provision was made for them with troughs and wells, but during the winter they were prone to freezing and much work had to be carried out in bitterly cold temperatures to ensure they were watered. Mules, evidently, were extremely picky about their drinking water and would never partake of muddy water.

The REs worked to install pumps and lay pipes to areas where water was needed, such as in the winter of 1917 when a pump was installed at Suzanne. Unfortunately, the pipes could not be buried very deep owing to the frozen ground, but, even in those very trying conditions, they were successful.

Reassembling a gun at a repair shop. These shops were used for major overhauls and repairs beyond the scope of the mobile workshops.

ANIMALS

The Army Veterinary Corps was well established, primarily for the treatment of horses. In August 1914 it consisted of 122 officers and 797 ORs looking after over 50,000 horses; these numbers rose with time.

Horses varied in breed depending on the task allotted to them, be it for officers' use, or heavy horses for pulling artillery pieces. It is a sad fact that the horses and mules were often casualties of shell bursts as it was nigh on impossible to give them protection when working in the open, and cavalry horses were quite literally mown down by machine guns during charges; they were also subjected to the effects of gas, and disease. This said, the work of the AVC was remarkable; the horses they were able to treat had a success rate of 80 per cent by the end of the war.

The AVC had sixty mobile sections and twenty horse hospitals.

Equines were not the only members of the animal kingdom represented on the Western Front. Carrier pigeons were extensively used with great success when all other means of communication were unavailable. Dogs were also used for carrying messages, but mostly kept for guarding and human comfort. Cats were there as pets, but were useful for keeping the rat population down, as were terrier dogs!

▶ A Thornycroft mobile artillery workshop attending to what looks to be an 18-pounder field gun. The gun has been stripped down with a dozen men working on it. Four more are in the workshop with the tools needed to deal with any repairs. A second lorry is alongside with stores for the workshop.

▼ A horse-drawn ammunition limber being loaded by some New Zealand soldiers.

Mice and canaries had a use for detecting gas in trenches and mines.

The RSPCA had temporary kennels at Boulogne for serving men and women to leave their dogs while they returned home for short leave. They also set up a fund to quarantine and repatriate dogs to Britain after the war.

AIR FORCE

There were numerous airfields on the Western Front used by the RNAS, RFC and RAF, but it was St Omer that demonstrated the evolution of the airfield. This, it has been said, was the spiritual home of the RFC. From a basic start in 1914 it evolved into what was described as a 'gigantic factory and emporium, repairing everything from aircraft to wireless equipment and

◀ A War Department Pacefield lorry loaded with timber at a timber yard, with labourers looking on from the sawmill beyond.

vehicles', and the stores included lawnmowers for keeping the runways trim. St Omer became an Air Park, with another opening at Candas in 1915, both being supplied by rail from the ports of Rouen and Boulogne. During the Spring Offensive of 1918, St Omer relocated to Guines, a little way out of Calais. By the end of the war there were seventy-seven squadrons with 1,645 pilots serving on the Western Front and 1,262 serviceable aircraft.

BALLOONS

The large kite balloons were used extensively behind the front line to observe the accuracy of artillery fire. The balloon was attached to a cable, which in turn was attached to a winch on the back of a 3-ton lorry. The winch speed depended on the wind speed; the higher the speed the slower the rate of descent, or haul-down, of anything from 900 to 450ft per minute. Owing to the change in pressure in the ear during the rapid descent, the observers were advised to keep their mouths open in an attempt to minimise any resultant deafness.

While suspended in a basket below the balloon, the men would relay their observations by telephone so that the artillery could be told if they needed to make any alterations.

The hydrogen for the balloons was originally obtained in France but growing consumption meant the building of hydrogen plants, one of which was at Arques, where kite balloon repairs were carried out.

◂ Timber stacked in a yard of a forest lumber works. Wood from local forests was processed at sawmills, often worked by German prisoners and transported around the site on a light railway.

▸ Duckboards being carried across a frozen canal. Duckboards were used as a causeway attached to an inverted A frame in the trenches, which helped to keep the men's feet out of the water.

GHQ

Originally the BEF General Headquarters (GHQ) was located at Blendecques, south of St Omer, as this was more central to the work of the BEF at that moment, but as time passed, and the BEF's area of control increased in Flanders, it was decided to relocate to the old fortified town of Montrueil-sur-Mer in March 1916, where it more or less completely took over the town. Situated out in the country away from any major town or base, it was only connected by a railway line, off the main coast line that ran from Étaples (7km distance away) to St Pol, on the south side of the River Canche. However, it was an ideal spot as it infrequently attracted the attention of enemy aircraft, plus it was centrally situated between Paris and London.

GHQ was 'the hub'. It was where everything was planned and organised. Many thought those who carried out work there had a cushy life, and certainly the living standards there would back this up, but it would hardly have been common sense to have had GHQ any closer to an area that was likely to have been destroyed by shells or bombing, and yes it was full of staff officers that arguably were out of touch with the realities of front line trench warfare, but as with all large organisations, be it military or commercial, there has to be somewhere where everything is co-ordinated.

The book *GHQ*, by Sir Frank Fox, relates all that went on at Montreuil, but here is a list of most of the various services controlled by the high-ranking individuals at GHQ: agricultural production, Army Postal Service,

The land shattered by shells near Zillebeke. Throughout the war anything that could be salvaged was, and here the devastated trees gouged by shrapnel are being pulled down and cut up. The army couldn't afford to waste this valuable commodity with its numerous uses.

➤ Water was vital for men, horses and vehicles. Here, petrol cans are being filled at a water point. Two tanks filled with water are connected by a long pipe, on which are a row of taps at intervals so a number of cans can be filled up together. The source of water could either have been sunken laid pipes or it could have come from ASC water tank lorries.

▼ A rare photo of No.10 Casualty Clearing Station at Remy (Lijssenthoek) when it first opened in 1915. In the beginning, water was stored in tanks as shown here, but by the following year, when the number of casualty clearing stations turned Remy into what resembled a small village, a more permanent water system was installed together with an electric light generating plant.

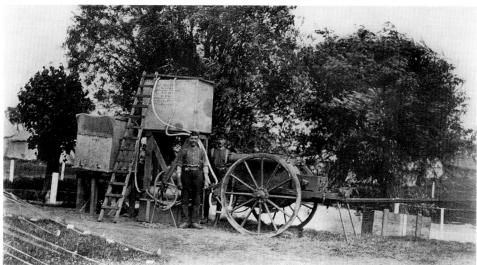

Expeditionary Force canteens, engineering stores, forestry, hiring, requisitions and claims, labour, remounts, pay-master, salvage, supplies, transportation, medical, veterinary, Imperial War Graves Commission (now CWGC), construction, docks, inland water transport (IWT), railways, roads, billeting, police, signals, traffic control, together with the various corps and brigades representatives, training, interpreters, intelligence, also maintaining good communication with town officials throughout the Western Front region, as well as the obvious continual communication with the British Government.

The Duval Barracks at Montreuil was originally a hospital for British and Indian troops prior to 1916 when GHQ took over. Douglas Haig's residence was

◀ Taking water from a hole in the ice wasn't without its problems. Although the ice and snow look clean and white it was often only after the ground had thawed that the men found something unsavoury in the area where they had earlier taken water.

just outside Montreuil at Beaurepaire, a country house south-east of Montreuil, at St Nicolas (on the D138 road). A mobile advance HQ train was built for him by the London and North Western Railway.

PICTURING THE FRONT

The government established the Press Bureau at the beginning of the war, under the auspices of the War Office. The first reporter tasked with informing those at home, although heavily censored, was Colonel Ernest Swinton who wrote under the name of 'Eyewitness'. Those at home wanted to know what was going on, and the government had to say something or they would have lost the trust of the populace. Although they were reluctant to use journalists, Swinton was joined by Henry Major Johnson of the *Daily News* later in 1914. Others followed: Philip Gibbs of the *Daily Telegraph*, Percival Phillips of the *Express*, William Beech Thomas of the *Mail* and *Mirror*, *Times* reporter Henry Perry Robinson and Herbert Russell of Reuters. All the reporters were subject to government control and their reports had to be submitted to C.E. Montague so they could be checked for content prior to publishing. Lieutenant Montague, *Manchester Guardian*, eventually went to Military Intelligence where he was involved with propaganda.

During the war, 35mm film wasn't available, so photographs were taken with cumbersome cameras, mostly needing tripods, as were movie cameras. As they were not conducive to taking 'action shots', many

▲ Once away from the trenches the men could have the luxury of a good shave, be it here in a dugout in the hill, or a shave and haircut later at a town or camp barber's.

▶ One of the reasons why Tommy wanted to get away from the trenches was to have a good clean. A wash is being taken in a bucket with water, as clean as possible given the circumstances. The soldier's trench-waders would have given him protection from deep water in the trenches.

of the photos were staged, leading to criticism. The official photos were issued by the Press Bureau or the Associated Illustration Agencies and 'circulated' by other agencies such as Central News and Topical News Agency, where newspapers and periodicals went to get the latest photo.

Much of the official photograph developing was done near to where the photo was taken, sometimes in a mobile laboratory, but wherever the processing was carried out it was important to have clean water to rinse the plates. Unfortunately, there were areas where there was little or no clean water available, such as Ypres where the geology dictated that little of the lying water had been naturally filtered or flushed with new clean water, meaning that it had a high concentration of iron together with other chemicals that were the result of the casualties and detritus of war leaking into the water course. This has caused the original plates that are in storage to deteriorate.

A name familiar to those with an interest in First World War photography is Realistic Travels, a company that specialised in the production of stereoviews. This company was based in London and had been in existence since 1908, and the company president and main photographer was Charles Milton DeWitt Girdwood. Girdwood, born in 1878 in Oxford Onterio, was a film-maker and official photographer for the British and Indian governments. He studied in Michigan and started a photographic business at the beginning of 1900. He was in India at the outbreak of the war and travelled to

France with the Indian Army, but he wasn't authorised to follow them to the trenches, so he travelled on to London. At the beginning of 1915 he photographed the effects of bombing in the city, and in April he was allowed to photograph the Indian hospitals at Brighton and Bournemouth, but he was desperate to go to the front. He had to be very persuasive but eventually the powers that be allowed him to go for, what was supposed to be, two weeks in the summer of 1915.

He shot his first film on the 29 July – a staged re-enactment of a trench attack – and, although Girdwood was enthusiastic to continue filming, there was a question mark over copyright ownership as to whether it belonged to Girdwood or the government. Although the majority of his stereoviews were produced in large quantities in specific topical sets, there are a small number that are quite rare, a couple of which are reproduced in this book.

There were cameras available for amateurs. Kodak had the folding soldier's vest pocket camera that used the new 127mm film. The pictures taken with these were often used as postcards. With the autographic variant of the camera it was possible to open the back and write information with a stylus on the film. These amateur photographs are every bit as important as the official ones as part of the overall history.

It would be wrong to write about photographers and film-makers and exclude two important men: Geoffrey Malin and Frank Hurley. Malin, together with John McDowell, was famous for the propaganda film *The Somme*, released in August 1916. Malin made twenty-six films in total, often in dangerous conditions.

▲ The Royal Engineers formed a Camouflage Section at the beginning of 1916, although they had a small detachment before then. The men called for to work in this section were artists, craftsmen and those who made theatrical scenes. The art of camouflage wasn't just to deceive the eye from a distance with a screen, but also to conceal something or someone in an object like an artificial tree. The work of this section was very successful during the war.

◄ Camouflage didn't have to be elaborate. These horse teams are bringing up large branches to be used to cover the big guns so they merge into the natural surroundings.

This French village has been taken over by the cavalry having a well-earned break. It was important for men and horses to have rest, food and water, especially on a hot day like this one.

Australian Frank Hurley had already made a name for himself before the war with his photography in the Antarctic and he became official photographer in 1917. He took some of the most iconic shots of the war, but was often criticised for his composite photomurals that were not a factual record of events.

INFANTRY BASE DEPOTS

Fresh from England, the new infantry men would be held in camps where they would train and wait for a posting to a front-line unit. Ideally situated near a Channel port, the main camps were at Boulogne, Rouen, Le Havre and Calais, with the largest at Étaples.

Situated south of Boulogne, the fishing port of Étaples, at the mouth of the River Canche, was transformed into a garrison town during the war. It was not a place remembered with fondness by most of the infantry who found themselves there; the notorious Bull Ring, where training and discipline was, so it was reported, taught with inhumane brutality, led inevitably to tension between the soldiers and those in authority. This tension reached a peak on 7 September 1917 when New Zealand gunner A.J. Healy was arrested. Men gathered and refused to disperse until he was set free. The situation was made worse when the military police arrived, and one, Private Reeve, fired at the crowd, wounding a French woman and killing 21-year-old Corporal Wood of the Gordon Highlanders. The military police retreated into the town followed by an angry mob.

With further demonstrations occurring, reinforcements were called for, and the situation was finally calmed on 12 September. It is hard to imagine what it was like at Étaples, but the most telling thing was that many a man who had spent two weeks there said they would rather return to the front.

With over twenty hospitals, Étaples had the highest number of hospitals on the Western Front. They included British, Canadian and St John Ambulance stationary and general hospitals, which were situated on the eastern side of the railway with their own siding. Above these, at the top of the hill, were war dog kennels, gas training areas, a mortuary and cemetery; now the largest CWGC cemetery in France. The camp

OTHER ACTIVITIES BEHIND THE FRONT 73

◀ Men of the 15th (The Kings) Hussars at a rest camp near Ypres with their horses. Horses were vulnerable during bombing or shelling as they were tethered, and the men had to do all they could to calm the animals and move them, if they could, to a safer area at great risk to themselves.

▶ The signals room General Headquarters was equipped with all the machines necessary to keep up with the flow of information both inwards and outwards. This room was in stark contrast to the conditions encountered towards the front, with lines being constantly shelled and repaired, more often under fire, with some sections of the line being destroyed with monotonous regularity.

▼ French farm buildings are being used as an advance remount depot winter quarters in 1915. The smiths wait with an anvil to carry on with their work of shoeing the horses.

and hospitals were subjected to air attacks in May 1918, with great loss of life. The bridge over the river was also damaged.

While at camp, men welcomed the opportunity to clean themselves and their uniforms, and delousing was carried out with clothes boiled or steamed in an attempt to rid them of lice. Horse-drawn steamers enabled clothes to be cleaned where there were no facilities available.

Camps and billets were found in farms, villages and towns near to the front, such as the large number of camps around Poperinge. Here the men could relax after returning from the trenches, or wait anxiously to go to the trenches.

◀ Being in a balloon was by no means a comfortable occupation. They were easy targets for artillery and air attacks, and, unlike their comrades in the air forces, they were equipped with parachutes for a quick escape. Here, the parachute is being inspected for any defects. Warm gloves and leggings seen here were essential for high-altitude work.

◀ A petrol-powered air-balloon winch mounted on the back of a Leyland truck.

▶ Taken on 20 September 1918 by the Royal Air Force, this aerial photo shows the area just over the Belgian border at Neuve Église (Nieuwkerke), to the north-east of Ploegsteert. Taken in the early morning, by the long shadows pointing west, means the buildings, trenches and shell holls are well defined. On the main road running to the top of the photo (Messines Road) can be seen farm buildings, some of which were used as camps with easy access to the trenches which can be seen across the top quarter of the photo. The largest building on the left of the road was Emu Farm Camp.

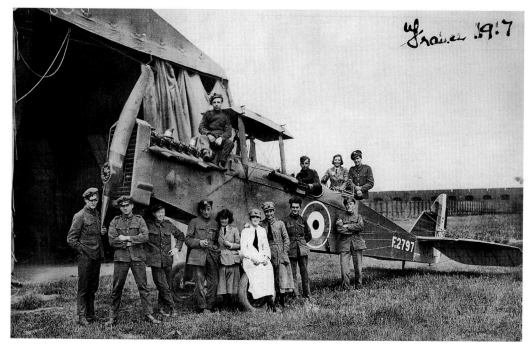

◀ This photograph was taken at an aerodrome in France in 1917 and shows the riggers and fitters outside a hangar. With the men are four women in uniform, but they are not of the Women's Royal Air Force as, like the Women's Royal Naval Service, it didn't come into being until the following year. It is possible they might be members of the Women's Army Auxiliary Corps, although their uniforms are remarkably similar to that of the Women's Royal Air Force. It is difficult to make out any cap badges, but it was not unusual for the women to adopt the badges of the unit they were attached to.

▼ Marshall Sir John French is pictured here with his ADCs at General Headquarters.

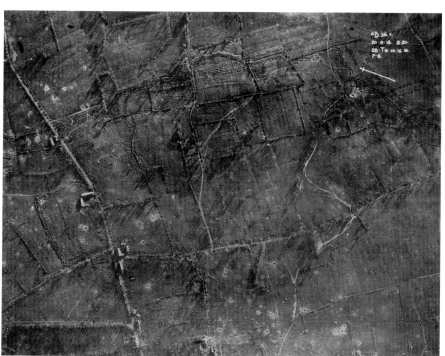

OTHER ACTIVITIES BEHIND THE FRONT 77

▲ There were regimental bands, but here we see the General Headquarters Troops Orchestra BEF with the musicians and instruments outside on a freezing winter day in January 1917 at Montreuil.

◀ Located at the entrance to the telecommunications centre at General Headquarters was the Scottish Churches Hut, where Haig would attend the Sunday service.

▲ The Prince of Wales is having a demonstration in the art of film-making by Charles Girdwood, who was president of Realistic Travels and produced the well-known series of stereoviews.

▲ Men take the opportunity to have a rest from their work to watch the official war artist Muirhead Bone while he sketches one of his famous scenes.

◄ An illustration by Fergus Mackain. His *Sketches of Tommy's Life* were extremely popular with the troops, and sold well, so well that he produced four series of ten cards, which were published in Paris and Boulogne. Here he takes a wry look at life in a country billet.

OTHER ACTIVITIES BEHIND THE FRONT

◀ British and Indian officers of the 1/1st Gurkhas at their headquarters in a village in Flanders. It must have taken the Indian troops some time to acclimatise to living on the Western Front, especially those who arrived in November 1914 to be confronted by a European winter in trenches at Neuve Chapelle in preparation for the battle the following spring.

▶ Men of a highland regiment building a road in their camp, which looks to be an extension of an existing camp as there are Nissen huts already in place and tents on the skyline. The Highlanders food store is laid out: crates of food, sacks of what look like turnips and loaves of bread. The men are wearing 'Tam O'Shanter' hats and most with kilts have khaki aprons that helped reduce the amount of dirt and water adsorbed into the woollen cloth, which was an ideal place for harbouring lice!

◀ This is the army camp at Boulogne, with its corrugated-iron huts and tents. The neat paths and gardens give the place an air of permanence. Behind the camp was a Marconi radio station.

▼ The 1/4th Loyal North Lancs Regiment in 1916 getting ready near their billets before going to the trenches. Surrounded by all their equipment, this photo is fascinating because although the photographer has asked them all to cheer, a close inspection of the men shows them all to appear genuinely at ease with hardly an anxious expression to be seen, which is remarkable given where they were going and the coming action, and the fact many would not return.

➤ Communal washing facilities at a camp in France. The long table arrangement has a ledge attached at the base for the men to stand on to keep them off the ground and stop it becoming muddy.

➤ American-Canadian Peter Norman Nissen designed a hut suitable for housing troops, or for use as a hospital ward, in April 1916. The shape of the hut, with its corrugated-iron roof, was designed to deflect shrapnel and reduce the effects of bomb blast. They were lightweight, easy to transport by rail and lorry, quick to assemble by six men in four hours and had many advantages over tents. Nissen conceived the idea while serving as a captain in the Royal Engineers and was later promoted to major. He is seen here talking to a non-commissioned officer.

▼ Playing cards was a popular leisure pursuit when the men were at rest and although there was a great temptation to gamble, non-commissioned officers and police would keep a vigilant eye on the men.

Gas was a greatly feared weapon, and gas schools were established to make the soldiers aware of its dangers and what could be done to combat its effects. Here men of the Guards are having their respirators inspected.

▲ A recreation hall somewhere in France serving non-alcoholic drinks, presumably Bovril, and snacks. It might not look much, but it was somewhere to relax, play games and forget for a while. This card was posted in November 1915 via Army Post Office No. 3.

➤ The Young Men's Christian Association provided many canteens for the men on the Western Front, and here is mobile kitchen serving hot drinks and snacks for the men.

➤ Religion played a great part during the war, and all denominations were represented to cater for the various nationalities and sects within the Allied forces. Most of the soldiers found comfort in prayer, be it the simple soldiers' prayer before committing themselves to the fight, or during a service such as this one that was marking the commencement of the fourth year of the war.

OTHER ACTIVITIES BEHIND THE FRONT   85

▲ Outside the mayor's house in the newly captured Nesle. Now was the time for troops to relax, with the expectation of experiencing the pleasures of a new town, going about finding supplies they might want, or visiting the barber, but by the look of most of them there wasn't much on offer.

➤ This postcard is of a sketch of the Young Men's Christian Association hut at Étaples. This was a very popular place for the men at the camp, giving them a chance to relax, read and watch shows in what was otherwise a camp with very little for the men to cheer about.

▲ Here, behind the men, is a mobile steamer for cleaning uniforms. Uniforms were not just dirty but riddled with lice and the clothes were put into this steamer to try to eradicate the unwanted companions. The machine could be taken to wherever it was needed, pulled by horses.

➤ Sports were an important part of service life and could range from tug-of-war to football matches, but here some kind of water volleyball is being played in a canal, with spectators lining both banks

➤ When at rest the men looked forward to various forms of entertainment available. Here, the 'Balmorals' are giving a theatrical turn for their fellow colleagues of the Black Watch after a sports event had been held.

RELAXATION

When away from the front line it was an opportunity for all to relax and mentally escape, however briefly. Concert parties were held wherever they could, with some of the playing groups gaining notoriety.

There were great temptations while out of the line, and money saved while men were in the trenches was now available to spend, be it on simple things such as egg and chips at an *estaminet*, drink, gambling or a visit to a brothel; the officers visiting brothels with blue lights, while ORs had to make do with the establishments showing the more familiar red lights. There were 137 *maisons tolérées* in thirty-five towns known to the authorities, and providing they passed

OTHER ACTIVITIES BEHIND THE FRONT 87

medical inspections and there had been no rowdy behaviour they were not out of bounds, but there were also many unofficial brothels, and some 150,000 British contracted a sexual disease during the war.

Those of a more temperate nature would seek out less bawdy establishments such as the Young Men's Christian Association (YMCA) huts. The YMCA was responsible for the first recreation centres of the war in Britain, providing refreshments and somewhere to rest near railway stations and alike for travelling troops. From November 1914, centres were opened to serve the Western Front troops at Le Havre, Dieppe, Dunkirk, Calais, Boulogne, Étaples, Abancourt, Aire, Abbeville, together with numerous stall-type arrangements and mobile canteens established at the side of roads to cater for the troops in the rearward areas from the trenches.

The YMCA also opened a hostel for relatives who were visiting their loved ones in hospital who had little chance of surviving.

There was a Scottish lad, Ross Martin, who was disappointed that he couldn't enlist because of his age, but undaunted he went to Rouen and worked there in the YMCA hut. He finally got his wish and enlisted. He wanted to be in the RFC, but a weak heart meant he could not and so he enlisted in the Tank Corps. Second Lieutenant Martin died of wounds in 1918, aged 19. He had only been at the front a few days.

Some sought peace and spiritual sanctuary, and the most well known for this was the 'man's club' Talbot House, or Toc H at Poperinge (founded in 1915), and even today the tranquillity there makes it an oasis for travellers.

▲ The female carpenters of W.G. Tarrant's workshop Calais 1917–18. The women are constructing panels to make up a hut. Tarrant had designed moulds that were the exact design needed to produce the panels, and were fixed to the bench. The women had to lay the wood planks in the moulds and fix them together to build up a panel.

FEMALE CARPENTERS

In January 1917 the first group of twenty women began work 3 miles outside of Calais making component parts for the construction of wooden huts. The women were fully trained in the art of carpentry and worked for Walter George Tarrant. Tarrant, from Surrey, was a builder of superior homes, but had a contract with the War Office to supply timber huts to the BEF. Shortage of male workers at home encouraged Tarrant to use women, and as timber made better use of cargo space than completed huts it was decided that the construction work would be carried out in France.

Special workshops were built alongside a canal so timber could be delivered directly to the carpentry works and the finished hut panels could be loaded onto barges for shipment inland. Tarrant also provided a self-contained hutted village for his women with

> Tarrant based his carpentry workshops next to a canal on the outskirts of Calais so the components of the huts made by the female carpenters could be transported in 'flat pack' form to where they were needed. There appears to be civilian loaders working on the barge.

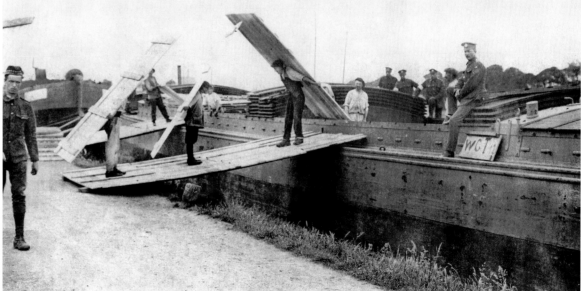

OTHER ACTIVITIES BEHIND THE FRONT

every convenience installed for their comfort, such as hot and cold running water, heating and a laundry. Tarrant also opened workshops for French women in another district of Calais and the English and French women together made 37,000 huts.

PRISONERS OF WAR

In total, 328,811 German prisoners were taken by the British in France. In the beginning the prisoners were housed in prison ships at Rouen and Le Havre, where they were put to work unloading vessels under guard. The ships were later replaced by ten prison camps located at ports and railheads where the prisoners could be usefully employed. In 1916, prisoners were taken to Abbeville where it was determined if any were suitable for the POW Labour Companies, especially if they had experience in engineering or forestry.

▲ Prisoners being held in a cage that was a compound to hold the men before being moved on.

▶ Thankfully today the Commonwealth War Graves Commission continues to care for the graves of the war dead in their beautiful and peaceful cemeteries like this one at Lijssenthoek.

▼ German prisoners being escorted by British soldiers near a railway. The number of men escorting outnumber the prisoners.

▼ An army chaplain attends to a grave, while a stretcher in the background and the shovel in the foreground indicate this is a new grave.

APPENDIX

RELEVANT PLACES

Abancourt	No. 1 Engineering stores depot/POW camp
Abbeville	Ambulance train depot and supply store/veterinary hospital/Third and Fourth Army regulating station
Aire	RE special works park
Arques (St Omer)	IWT workshops and dry dock/balloon sidings/ammunition dump
Audruicq	ROD workshop/RE store/ammunition dump
Bailleul (Belgian border)	Engineering shop and sawmill
Beaurainville	WDLR workshop (mid-1918)/ mechanical engineers workshop
Beauval	Omnibus workshop
Berguette	WDLR main workshop
Blarges	Ammunition depot
Blendecques (St Omer)	First location of GHQ
Bois de Cise (Le Treport)	Tank camp
Borre	ROD locomotive workshop
Bourbourg	Ammunition depot
Boulogne	Channel port/army postal depot/ROD depot/rest camp/bakery/ Marconi wireless station
Brandhoek	Petrol dump
Buire	ROD depot
Calais	Channel port/army postal depot/bakery/ROD depot/rest camp/army printing and stationery depot/ordnance clothing depot
Camiers	ROD depot
Dannes-Camiers	Ammunition depot
Dieppe	Channel port
Doullens	Railhead
Dunkirk	RNAS station/cross-Channel barge dock
Erin	Tank depot
Étaples	Base camp/numerous hospitals' training ground (Bullring)
Frévent	Rest camp
Gézaincourt	Railhead
Heilly	Railhead
Helfaut (St Omer)	RE Special Brigade gas laboratory
Hesdin	ASC tyre-pressing depot
Le Havre	Channel port/army postal depot/ No. 3 engineering store (base park)/canal depot/cookery school/bakery/rest camp/army printing and stationery depot
Le Lacquer	Light railway workshops
Le Rivesnay	Signalling school
Le Touquet	Claims commission
Lys Canal (Near Aire)	RE bridging school
Marquise	ROD depot/stone quarries
Martin Église (Dieppe)	Chalk quarry
Mers	Tank depot
Montreuil	GHQ
Noyelles	Chinese labour depot

The devastated landscape of the battlefield. Here items suitable for recycling, such as shell cases, would be collected. Searches would also be carried out to recover the war dead for burial.

Oissel	Transportation depot
Pont Remy (Abbeville)	Musketry school
Puchevillers	Railhead
Quevilly (Rouen)	ROD depot/petrol depot/hydrogen plant
Reninghelst	Australian workshops and gun repairs
Rouen	Docks/ASC No. 2 base motor transport and traction engine repair shop/No. 3 repair shop (omnibus)/railway works/RE base depot/ gas school/army printing and stationery depot
Rouxmesnil	Ammunition depot
Sanvic (Le Havre)	Rest camp
Serqueux	Sand mine

Soquence (Le Havre)	ROD camp
St Omer	Home of the RFC and RAF/ASC heavy repair shop
St Valery sur Somme	Rest camp/omnibus park
Teneur	Tank corps central workshops
Terlincthun	Railway cadre camp
Vequement	Railhead
Vron	RAF supply depot and store of 3 million gallons of petrol
Wardrecques	Ordnance Survey overseas branch, 1918
Wimereux	WAAC (QMAAC) HQ/ RE camouflage factory/ Ordnance Survey (moved here after Spring offensive, 1918)
Wisques	Machine-gun school
Zeneghem	Ammunition depot/ baulk timber store

BIBLIOGRAPHY

BOOKS
William A.T. Aves, *R.O.D. Railway Operating Division on the Western Front*, Shaun Tyas Publishing, 2009
Chris Batten, *Ambulances*, Shire Publishing, 1996
Peter Cooksley, *The RFC/RNAS Handbook 1914–18*, Sutton Publishing, 2000
W.J.K. Davies, *Light Railways of the First World War*, David & Charles, 1967
Martin & Joan Farebrother, *Tortillards of Artois*, The Oakwood Press, 2008
Sir Frank Fox, *G.H.Q.*, Philip Allan, 1920, Naval & Military Press reprint
Colonel A.M. Henniker, *Transportation on the Western Front*, Imperial War Museum & The Battery Press, 1992
T.R. Heritage, *Light Track from Arras*, Plateway Press, 2007
Institute of Royal Engineers, *History of the Corps of Royal Engineers*, 1952
Roy Larkin, *Destination Western Front*, Roy Larkin, 2010
Murray Maclean, *Farming and Forestry on the Western Front*, Old Pond Publishing, 2004
Philip Pacey, *Railways of the Baie de Somme*, The Oakwood Press, 2000
John H. Plumridge, *Hospital Ships and Ambulance Trains*, Seeley Service & Co., 1975
Statistics 1914–1920 of the Military Effort of the British Empire during the Great War, printed by the War Office in 1922, Naval & Military Press reprint
Andrew Rawson, *British Army Handbook 1914–1918*, Sutton Publishing, 2006
Neil Storey & Molly Housego, *Women in the First World War*, Shire Publishing, 2010
John Sutton, *Wait for the Wagon (Royal Corps of Transport)*, Leo Cooper, 1998
Keith Taylorson, *Narrow Gauge at War*, Plateway Press, 1987 & 2008
Keith Taylorson, *Narrow Gauge at War 2*, Plateway Press, 1996
Michael Young, *Army Service Corps 1902–1918*, Leo Cooper, 2000

WEBSITES
www.1914-1918.net – The Long, Long Trail
www.ramc-ww1.com – Royal Army Medical Corps in the First World War
www.remembrancetrails-northernfrance.com – Nord Pas de Calis Chemins de Memoire 14–18
www.scarletfinders.co.uk – British Military Nurses
www.theaerodrome.com

MAGAZINES
Britain at War
The Bulletin (The Western Front Association)
Stand To! (The Western Front Association)

PRIVATELY PRINTED LITERATURE
'No. 103 The supply of railway stores at home and with HM Forces in the Field', Great Western Railway Debating Society, 28 October 1920:
'The work done by railway troops in France 1914-19', Institute of Civil Engineers, 1920
Women carpenters in France by W.G. Tarrant – a letter to women's work sub-committee 1919

The Royal Engineers built bridges of all types, both temporary and permanent. This wooden bridge has stout uprights that have been driven deep into the river bed for stability.

INDEX

Abbeville 6, 9, 23, 88, 90, 92, 93
Ambulance Train 23–8, 54, 92
Ambulances 19, 20, 22–6
Armoured car 6, 32, 35, 37–8
Army Service Corps (ASC) 10–3, 16, 31, 35, 37, 39, 63, 69, 92–3
Arras 9, 42, 47, 49–51, 57
Audruicq 50, 53, 54, 56, 59, 60, 92
Balloons 66, 76, 92
Barges 19, 29–31, 88
Bicycles 39, 42, 80,
Boulogne 9–12, 23, 49, 50, 53, 59, 63, 65–6, 73, 79, 82, 88, 92
Bread 9–11, 13, 80
Calais 9–12, 23, 28, 35, 50, 56, 63, 66, 73, 88, 90, 92
Camp and billets 9–11, 13, 27, 33, 63, 69, 71, 73–4, 76, 79–83, 86, 92–3
Canals 19, 30–1, 67, 86, 88–9, 92
Carpenters, female 88–90
Coal 9, 12, 54, 63
Depots 9–10, 35, 49, 54, 60, 73–4, 92–3
Etaples 11, 23, 63, 73, 86, 88, 92
Filming and photography 70–3, 79, 79
Forestry and timber 5, 63, 66
General Headquarters (GHQ) 67, 69–70, 74–5, 77–9
Girdwood, Charles 71, 72, 79
Hazebrouck 9, 50
Horses 4, 5, 10, 13, 19, 21–2, 25, 36, 42–3, 45, 64–5, 72–4
Hospitals 23, 74
Le Havre 9, 11–2, 23, 49, 59, 63, 73, 88, 90, 92
Lorries 11, 31–3, 35, 38, 42, 44, 65–6

Maps 6, 42, 44, 49, 60–1
Motorcycles 33–6 39
Munitions 22, 51, 54, 55, 62–3
Omnibuses 11, 39–41, 44
Petrol 9, 33, 35, 40 92–3
Poperinge 40, 61, 76, 88
Postal service 12–3, 16–7
Press Bureau 70–1
Prisoners of war 90
Railway 9, 21, 48–61
Recycling 63, 68–9
Relaxation 85–8
RFC, RAF 32, 39, 65–6, 76–7, 88, 93
Richborough 60, 96
RNAS 22, 35, 37, 65, 92
Roads 19, 32–3, 38–9, 41–5, 47, 56, 58–9, 63, 69
Rouen 6, 11–2, 19, 23, 31, 35, 39, 52, 63, 73, 88, 90, 93
Royal Engineers 12, 30, 45–7, 49–61, 72, 83
Schools 11, 84, 92–3
Somme 9, 11, 27, 49, 52, 72
St Omer 63, 65–7, 92
Tractors 33, 40, 52
Tanks 59–61
Water 13, 24–7, 35, 58, 63, 69–71, 73
Women 13, 20, 21, 25, 34, 49, 65, 77, 88–90
Workshops 33, 38–40, 52, 53, 59–61, 64–5, 88–9, 90, 92–3
YMCA 85–6, 88,
Ypres 6, 9, 20, 39, 45, 50, 52, 58, 71, 74

Visit our website and discover thousands of other History Press books.

www.thehistorypress.co.uk

So busy was Dover with the cross-Channel barges that the port was overcrowded and an alternative port had to be found. Richborough, further up the Kent coast, was chosen. Here is ferry No. 3, on 29 December 1917, with wagons full of railway supplies and munitions boxes ready to start one of the first ferry trips.